Water Circulation in Rocks

Laura Scesi · Paola Gattinoni

Water Circulation in Rocks

Laura Scesi
Full Professor
Piazza Leonardo da
Vinci, 32
20133 Milano
Politecnico di Milano
Dept. of Environmental, Hydraulic
Infrastructures and Surveying Engineering
Italy
laura.scesi@polimi.it

Paola Gattinoni
Assistant Professor
Piazza Leonardo da
Vinci, 32
20133 Milano
Politecnico di Milano
Dept. of Environmental, Hydraulic
Infrastructures and Surveying Engineering
Italy
paola.gattinoni@polimi.it

The Italian language edition of the book La circolazione idrica negli ammassi rocciosi was first published by Casa Editrice Ambrosiana in 2007.

ISBN 978-90-481-2416-9 e-ISBN 978-90-481-2417-6
DOI 10.1007/978-90-481-2417-6
Springer Dordrecht Heidelberg London New York

Library of Congress Control Number: 2009926836

Cover illustration: Cover image © JupiterImages Corporation

Printed on acid-free paper

Springer is part of Springer Science+Business Media (www.springer.com)

Contents

Chapter 1
Introduction to Water Circulation in Rocks

1.1 General Observations

Water circulation in rocks is a very important element in the solution of quite typical problems arising in environmental, civil and mining engineering. Still, aquifers in rocks are devoted much less attention than those in porous media, partly because they are considered less important from the point of view of water research, and partly because rocks are a very complex medium; therefore its modelization is quite complex. Actually, the presence of fractures plays a fundamental role in the hydrogeological characterization of rock masses; it determines an increase in hydraulic conductivity of different magnitude along the main direction of the fractures. A further aspect, fundamental but very difficult to represent, is the presence of local strips of rock alteration on the land surface that strongly influence both the infiltration process following meteorological events and the supply of deeper aquifers.

Water circulation in rocks occurs through a system of "vacuums" that is quite different from that of soils in dimensions, shape and density. In most rock masses, water circulation occurs through the many primary discontinuities (stratification, schistosity, karstic cavities) and/or secondary discontinuities (fractures, faults, karstic cavities).

In *intrusive rocks*, for example, vacuums are mainly represented by fractures; in *metamorphic rocks* by fractures and schistosity planes; in *sedimentary rocks,* vacuums are often determined by dissolution that widens already existing discontinuities (karsts phenomena), by the lack of cementation of part of the rock and by fracturing or stratification. Tuff shows cavities that are caused by the disintegration of ashes or the dissolution of limestone present in the original rock, by a dishomogeneous consolidation of the mass or by the lack of cementation of part of the rock. Porphyries, as well as basalts, show a marked columnar cracking due to the volume decrease of rock during the cooling process. In lavas and volcanic scoriae, the vacuums partly caused by "degassing" while cooling give origin to an extremely high permeability and an exceptional porosity, such that they create relevant water accumulations. These types of vacuum allow water circulation in rocks and also in the ones that are little or not permeable at all due to their lithologic nature (Fig. 1.1 and Table 1.1).

L. Scesi, P. Gattinoni, *Water Circulation in Rocks*, DOI 10.1007/978-90-481-2417-6_1,

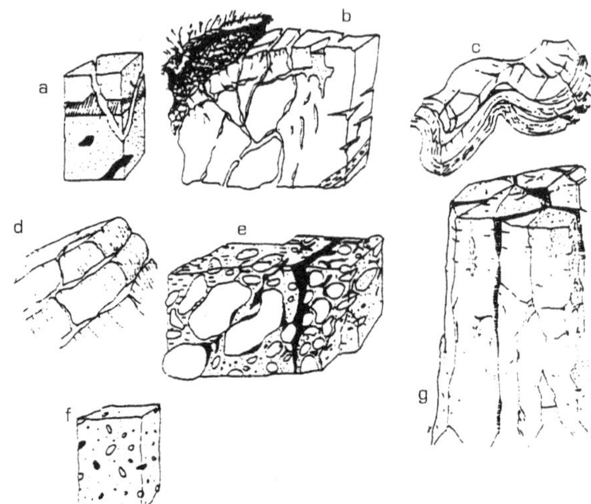

Fig. 1.1 Types of vacuums typical of rock masses (Francani, 1997): (**a**) Fractures in intrusive rocks; (**b**) leptoclases in weathered superficial rocks; (**c**) schistosity and fractures in metamorphic rocks; (**d**) karstic dissolution in carbonatic rocks; (**e**) vacuums due to lack of cementation, fractures or karstic dissolution in rubbles and conglomerates; (**f**) syngenetic porosity; and (**g**) columnar cracking

Table 1.1 Comparison between porosity and permeability values (increasing in the direction of the arrow) of some types of rocks (Civita, 2005)

Porosity	Permeability
Intact crystalline rocks	Intact crystalline rocks
Fractured crystalline rocks	Clay
Karstic limestone	Tuff
Volcanic slags	Silt and organic silt
Sandstone	Fractured crystalline rocks
Fissured limestone	Sandstone
Uncemented tuff	Silty sands
Heterogeneous sand and gravel	Bioclastic and fractured limestone
Gravel	Heterogeneous sand and gravel
Well-grated sand	Well-grated sand
Silt and also organic silt	Gypsum and vacuolar dolostones
Organic clay	Porous volcanic rocks
Organic clay and peat	Well-grated gravel
Clayey mud	Karstic limestone

The water flow takes place inside this complex network of vacuums and it is strongly influenced by meteorological recharge, by the regime of superficial water streams and by the melting of snow. Therefore, knowing the origin, features and distribution of those vacuums is fundamental to better understand how water flows in rocks.

1.2 Origin of Discontinuities

Based on what was stated above, the "vacuums" present in rocks can be divided in two wide categories:

- Vacuums due to lack of cementation, degassing, dishomogeneous consolidation of the rock mass, dissolution of rock portions, etc.
- Vacuums generated by cooling phenomena, stratification, schistosity, fractures and faults of tectonic origins, karsts phenomena etc.

In the first case, the rock is divided in small fragments, similar to granules of soil; therefore, if there vacuums are interconnected, the water flow patterns are similar to those in porous media.

In the second case, the cracks network (discontinuities) divides the rock in quite big portions and the water flow results strongly influenced by the geometric and mechanical characteristics of those vacuums. Therefore, a distinction has to be made between the following:

- *Intact rock*, element constituted by granules or crystals, bound by permanent cohesive forces, with no discontinuities;
- *Rock mass*, physical body constituted by blocks of intact rock separated by discontinuities.

Intact rock is a continuous medium with almost no permeability and porosity, whereas rock mass is a non-continuous medium where discontinuities determine the hydraulic behaviours of the whole (Table 1.2). Therefore, it is important to know the features of that discontinuities network.

Table 1.2 Types of vacuums (modified by Civita, 2005)

Scale	Type of vacuums	Type of medium	
Microscopic (< 1 mm)	Pores	Porous	Continuous
	Microfessures	Fissured	Discontinuous
Macroscopic (> 1 mm)	Macrofessures		

1.3 Features of Discontinuities

Considering the peculiar features of water flow in rock masses, it is fundamental to carry out a geological-structural and a geological-technical study to highlight the main parameters and the main features of primary discontinuities (cooling cracks, stratification, schistosity) and secondary discontinuities (fractures and faults of tectonic origin, karsts dissolution).

1.3.1 Orientation

Discontinuities in rock masses can be grouped in families according to their *orientation* that constitutes the neighbourhood of a modal value typical for each family or set (Figs. 1.2 and 1.3) even though it shows high dispersity. The orientation of a discontinuity set in space is defined through the three angles that follow (Fig. 1.4):

- *dip direction*: horizontal trace of a the line of dip, measured clockwise from North;
- *dip*: the angle that the plane forms with the horizontal;
- *strike*: orientation, with regards to the cardinal points, of the line of intersection of the plane with a horizontal plane; it is always perpendicular to dip direction.

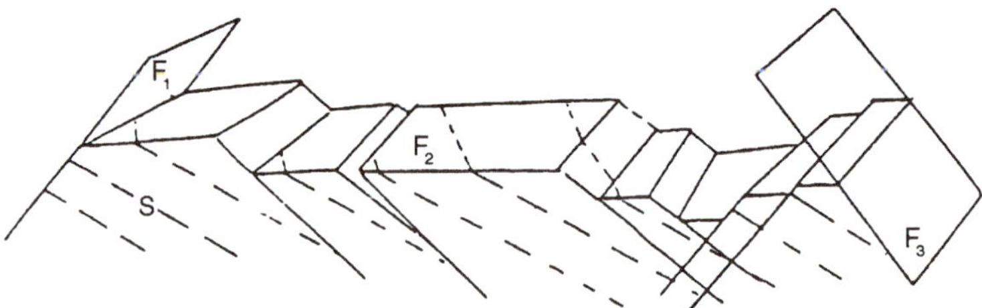

Fig. 1.2 Schematic representation of a rock mass characterized by the presence of four discontinuity families (S, F_1, F_2, F_3)

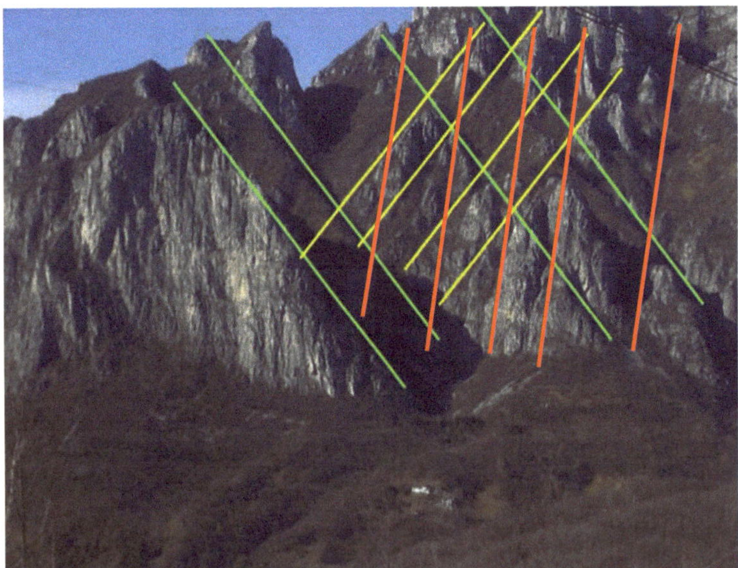

Fig. 1.3 Example of a rock mass made by intact rock and discontinuities; the three discontinuities families highlighted in *yellow*, *red* and *green* can be easily recognized (Scesi et al., 2006, modified)

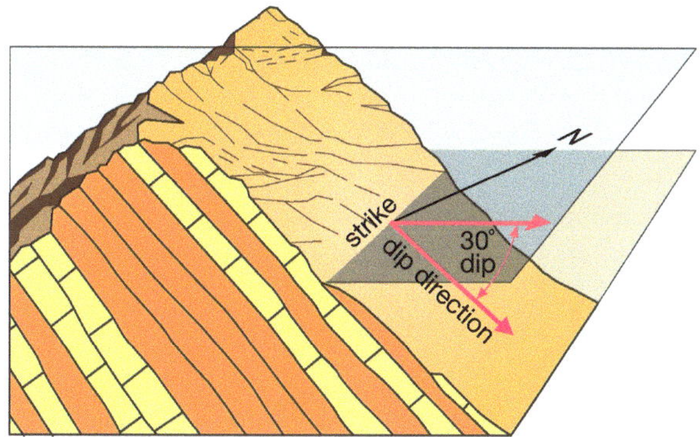

Fig. 1.4 Orientation of a plane in space

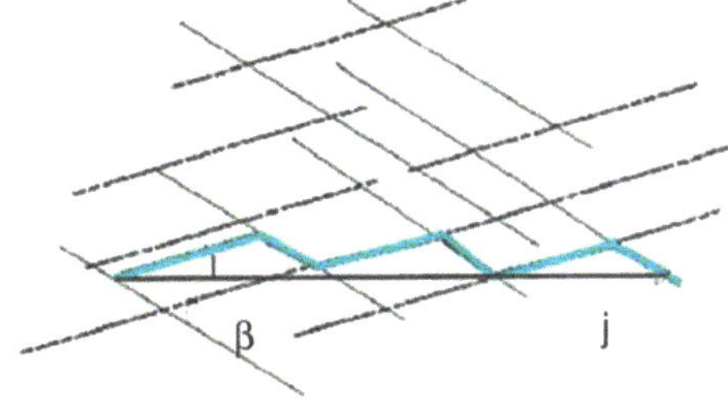

Fig. 1.5 Direction of the water flow inside a discontinuity network with a piezometric gradient j. The trajectory of fluid particles is conditioned by the hydraulic gradient, the permeability distribution and the discontinuity orientation (Francani, 2002)

In rocks with fissural or fracture permeability, water flows along confined paths (Fig. 1.5), thus determining a marked orientation of the water flow. As a consequence, it is evident that the discontinuities orientation strongly influences the flow direction; actually

- if discontinuity planes lie orthogonally to the groundwater piezometric gradient, the water flow is prevented, therefore permeability along that direction is null (Fig. 1.6a);
- if the discontinuities orientation occurs on planes parallel to the piezometric gradient, the water flow is maximum and the apparent permeability coincides with the effective permeability of the discontinuities (Fig. 1.6b);
- if the orientation of discontinuities is oblique to the piezometric gradient, the situation is intermediate; the closer the gradient is to the perpendicularity with regard to the discontinuity planes, the more the fluid flow is hindered, whereas the closer it gets to the orientation of the planes, the more the fluid flow is favoured. In any case, the real flow of the groundwater is diverted by the walls delimiting the discontinuities and liquid particles are bound to follow a longer path (Fig. 1.6c).

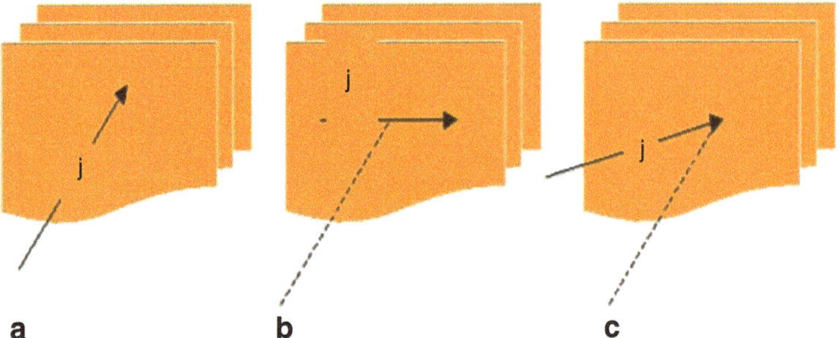

Fig. 1.6 Possible trends of hydraulic gradient with respect to discontinuities orientation: (**a**) The gradient is orthogonal to discontinuities planes; (**b**) the gradient is parallel; and (**c**) the gradient is obliquous

1.3.2 Degree of Fracturing

Quantity and proximity of fractures pervading a rock mass define the degree of fracturing of the rock (Fig. 1.7) and, as a consequence, its permeability. Therefore, if the areas with higher water flow must be detected, it is necessary to localize all those areas where the degree of fracturing is particularly high, e.g. the "shear zones" considered as high deformation areas with limited thickness with respect to their longitudinal extension (Ramsay and Huber, 1987). The in situ identification of these shear zones can become fundamental to identify the areas with the higher potential water flow, thus defining the main flow direction, delimiting the hydrogeological basin and identifying the recharge and discharge areas. Spacing, intercept, frequency, RQD and unitary rock volume are the features that allow the description of the degree of fracturing of a rock mass.

The spacing of discontinuities is defined as the mean distance among discontinuities belonging to the same family, measured perpendicularly to the discontinuities (Fig 1.8b).

The intercept is measured without considering the belonging of discontinuities to the different families and it represents the mean distance of discontinuities with respect to a base of measure.

The inverse of intercept, that is the number of discontinuities per meter, is showed as frequency or intensity of fracturing (Fig. 1.9). An empiric estimate of the degree of fracturing of a rock is provided by the RQD coefficient (Rock Quality Designation), representing the modified core recovery percentage of a drilling test and is given by the ratio between the sum of the pieces of the core with length over 10 cm and the core run total length. When no drilling tests are available, the RQD can be assessed in function of the frequency of fracturing using following empirical relations (Priest and Hudson, 1976):

$$\mathrm{RQD} = (115 - 3.3 \times \mathrm{J_v})\ (\mathrm{J_v} = \text{number of discontinuities per unit of volume}),$$

$$\mathrm{RQD} = 100(0.1\mathrm{f} + 1)^{-0.1f}(\mathrm{f} = \text{number of discontinuities per meter or frequency}).$$

Fig. 1.7 (**a**) Example of very fractured rock mass; (**b**) example of unitary rock volume (URV); (**c**) and (**d**) rock outcrops characterized by different fracturing degree

From the hydraulic point of view, a low value of spacing or intercept and, as a consequence, a high value of frequency favours underground water circulation, as they indicate a marked fracturing of the rock.

The rock mass, divided by discontinuities, is made of discrete elements of intact rock, indicated as *unitary rock volumes* (URV; see Fig. 1.7).

Generally, unitary volumes over 500 dm^3 are considered high, those in between 500 and 10 dm^3 medium, in between 10 and 1 dm^3 low, lesser than 1 dm^3 very low, corresponding to highly fractured rock. Evidently, the smaller the dimensions of unitary block, the bigger the quantity of water that can percolate.

Fig. 1.8 (**a**) Example of rock mass with thin and persistence stratification and (**b**) example of spacing between two discontinuities

Fig. 1.9 Example of open discontinuities characterized by variable spacing

The fracturing of a rock mass is generally variable in space, also in the function of the stresses to which the rock was subjected. As a consequence, the modalities of water circulation, in particular, in a carbonatic rock mass, show different peculiarities in different areas of the same mass (Fig. 1.10), also due to the permeability variations when depth changes, as it was experimentally observed by various authors (Snow, 1970; Louis, 1974; Gangi, 1978; Walsh, 1981).

1.3.3 Persistence

Persistence is defined as the extension of a discontinuity with respect to a reference line belonging to the plane on which the discontinuity lies (expressed as a percentage). This is the *linear persistence*. As a discontinuity surface can be formed by

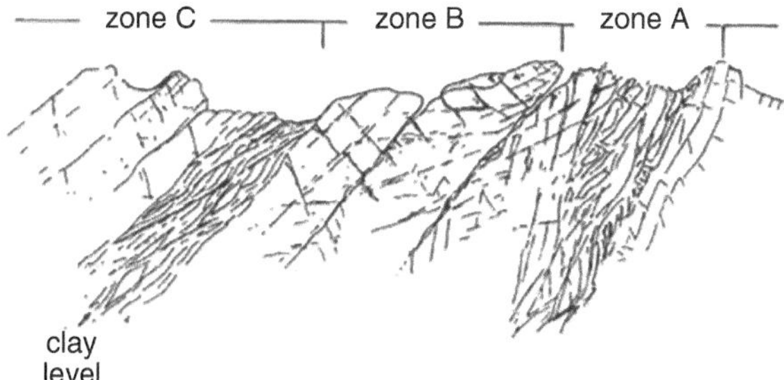

Fig. 1.10 Distinction among variable permeability zones, according to the fracturing degree of the rock. Groundwater flow occurs preferentially in the cataclastic zone (zone A); water flow along fractures prevails in zone B, whereas permeability decreases in zone C due to the presence of a clay level and because primary discontinuities prevail on fracturing (Francani, 1997)

zones presenting a total separation between walls and/or rock bridges that can have hydraulic characteristic similar to those of intact rock, the persistence of discontinuities can also be defined as the percentage ratio between the zone of effective separation and the area of the plane presenting the discontinuity (*areal persistence*).

Persistence is quite difficult to measure, therefore it is often evaluated as the discontinuity length on the outcrop. Generally

- if persistence is over 80% (see Fig. 1.8a), the hydraulic behaviour of the rock mass is basically conditioned by water flow inside discontinuities;
- if persistence is less than 25%, water flow is almost hindered due to the lack of interconnections among different discontinuities, which therefore create isolated and localized zones where water flows.

1.3.4 Aperture and Filling

The *aperture* of a discontinuity represents the distance between the discontinuity walls (Fig. 1.11); the discontinuity apertures are measured by mean of a thickness gage or a caliper and they are generally classified according to their size (Table 1.3). Outcropping apertures can be influenced by external factors, such as stress, loosening and superficial weathering, and they are usually wider than those present inside the rock mass.

In particular, discontinuities can be

- tight (rock–rock contact);
- open without filling;
- open with filling.

Fig. 1.11 Examples of open discontinuities, with no filling

Table 1.3 Classification of the discontinuity aperture (Barton, 1973)

Aperture (mm)	Classification
<0.1	Very tight
0.10–0.25	Tight
0.25–0.50	Partly open
0.50–2.50	Open
2.50–10.00	Moderately wide
>10	Wide

The presence of filling in discontinuities must be recorded considering its width, mineralogical composition (calcite, silt, clay, sand, etc.), grains size distribution and moisture conditions.

1.3.5 Roughness

Discontinuity surfaces can be planar, undulating, stepped or irregular (Fig. 1.12a). Roughness is the mean height of asperities of the surface relative to the mean aperture. It can be assessed in different ways:

- by direct contact of the joint surface, by mechanical profilographs (Fig. 1.13a; Barton and Choubey, 1977), or electronic profilographs followed by visual comparison (Beer et al., 2002);
- without direct contact with the joint surface, by photogrammetric techniques, interferometry, optic fibres, laser scanning etc. (Feng et al., 2003) and rielaborations that follow (Yang et al., 2001; Belem et al., 2000)

Some of these techniques can be only used on lab samples; others need expensive and sophisticated equipment but, in both cases, the assessment of asperities heights is always a complex procedure, very difficult to carry out in situ.

Fig. 1.12 (**a**) Examples of very undulated joints and (**b**) example of thickly stratified rock

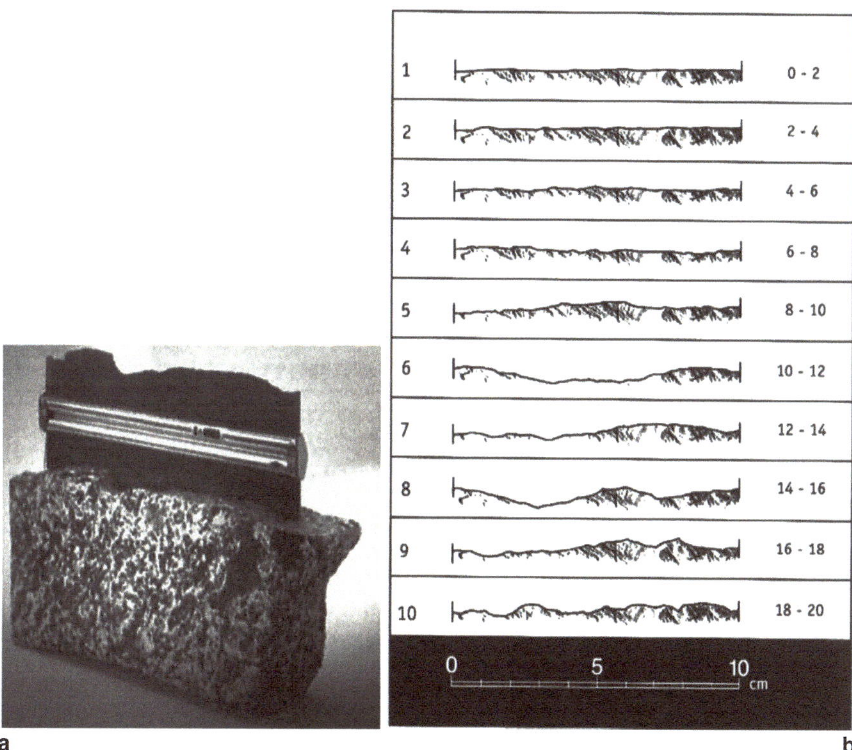

Fig. 1.13 (**a**) Shape tracer to measure the joint roughness and (**b**) typical roughness profiles for JRC range (on the right) (Barton and Choubey, 1977)

In geomechanical applications, Barton and Choubey (1977) introduce a *joint roughness coefficient*, called JRC, that can get around that difficulty. The values of the JRC were obtained empirically starting from shear resistance tests:

$$JRC = \frac{arctg(\frac{\tau'}{\sigma'_n}) - \varphi_r}{\log_{10}(\frac{JCS}{\sigma'_n})},$$

where

- τ' = peak shear strength along the joint;
- φ_r = residual friction angle;
- JCS = joint wall compressive strength;
- σ'_n = effective normal stress on the joint plane.

On the basis of empirical checks, the Author associated a series of standard profiles (Fig. 1.13b) to the numeric coefficient representative of roughness (JRC) that make use of that parameter particularly easily, notwithstanding the limits linked to the subjectiveness of the comparison of the profiles measured in situ using the shape tracer with the standard ones (Beer et al., 2002) and the fact that the tridimensional geometry of the joint is not taken into consideration (Grasselli and Egger, 2003).

From the hydraulic point of view, roughness reduces the effective aperture of discontinuities and its influence increases as the aperture gets smaller (Bandis et al., 1985). Various studies were carried out measuring how roughness and fracture apertures variations affect water flow (Brown et al., 1986; Gentier and Billaux, 1989; Piggott, 1990; Lapcevic et al., 1990); they highlighted that roughness and undulations of the discontinuities walls determine points of contacts and rock bridges that can be modelled through various statistical distributions of joints apertures. In the presence of rough joints, the application of a load can also considerably affect the permeability values of the joint; actually, dilatancy phenomena determine variations of the hydraulic aperture of the joints that are strictly correlated to those of the mechanical aperture (Lee and Cho, 2002).

1.3.6 Weathering

Joint weathering (Fig. 1.14) can either be measured directly in situ using a sclerometer or described using Table 1.4.

Evidently, the weathering process determines an increase of porosity and, therefore, a higher propensity of the rock to let water pass through; exceptions are represented by those cases where the weathering process leads to the formation of levels and clayey filling characterized by very low permeability.

1.3.7 Moisture Conditions and Seepage

As in most rock masses water flows inside discontinuities, it is important to differentiate

Fig. 1.14 Example of weathered discontinuity

Table 1.4 Level of weathering of joints surfaces

Degree	Description	Abbrevation
Not weathered	No visible traces of weathering on the discontinuity surface	WD1
Slightly weathered	The original colour of the rock partly or completely has changed on the discontinuity surface	WD2
Weathered	The weathering has not only interested the discontinuity surface but also the rock for about a millimetre. The original texture is still recognizable	WD3
Very weathered	The alteration is some millimetres deep with the complete transformation of the rock in soil	WD4

- tight, dry discontinuities, where water flow is hindered;
- open, dry discontinuities, with or without water flow;
- open, wet discontinuities;
- open discontinuities, where water flow is either continuous or discontinuous.

1.4 Graphical Representation of Discontinuities

Stereographic projections are used to represent data about discontinuities in rock masses in synthetic ways, as they allow a tridimensional view of the planar problem. To obtain those representations, let us imagine that we have a reference sphere whose centre is the origin of a Cartesian axes system; if a plane passes through the centre of the sphere, the intersection of the plane with the sphere creates a "great circle" that defines dip and dip direction of the plane in space (Fig. 1.15). This

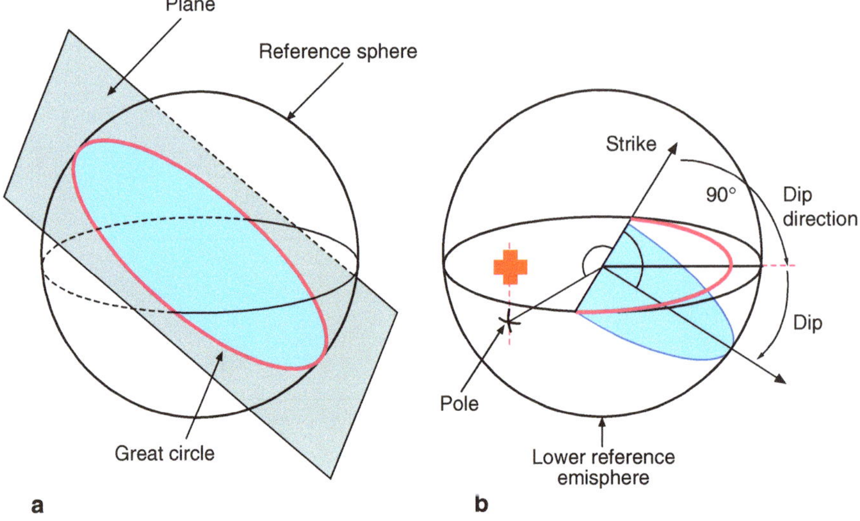

Fig. 1.15 Creation of stereographic projections

information is obtained observing both the upper and the lower hemisphere; for convenience, only one of the hemispheres can be used (generally the lower one). The same plane can also be represented by a point called a "pole", and not just by a "great circle". It represents the intersection of the straight line passing through the origin and perpendicular to the plane with the sphere surface.

The bidimensional representation of the large circle and of the pole is obtained projecting the reference sphere on a horizontal plane (*polar projection*) or on a vertical plane (*equatorial projection*). Generally, to have a clearer drawing, discontinuities are represented just with poles and not with great circles.

1.4.1 Equal Areal Projections

Schmidt's equal areal projections (the areas are kept constant; Fig. 1.16a) are used when the "distribution" of discontinuity orientation (or of other geological structures) must be known. This type of representation allows the grouping of poles according to their frequency percentage; in this way, those areas are obtained whose maximum pole concentrations are the pole of the mean planes of the discontinuity families (Fig. 1.17).

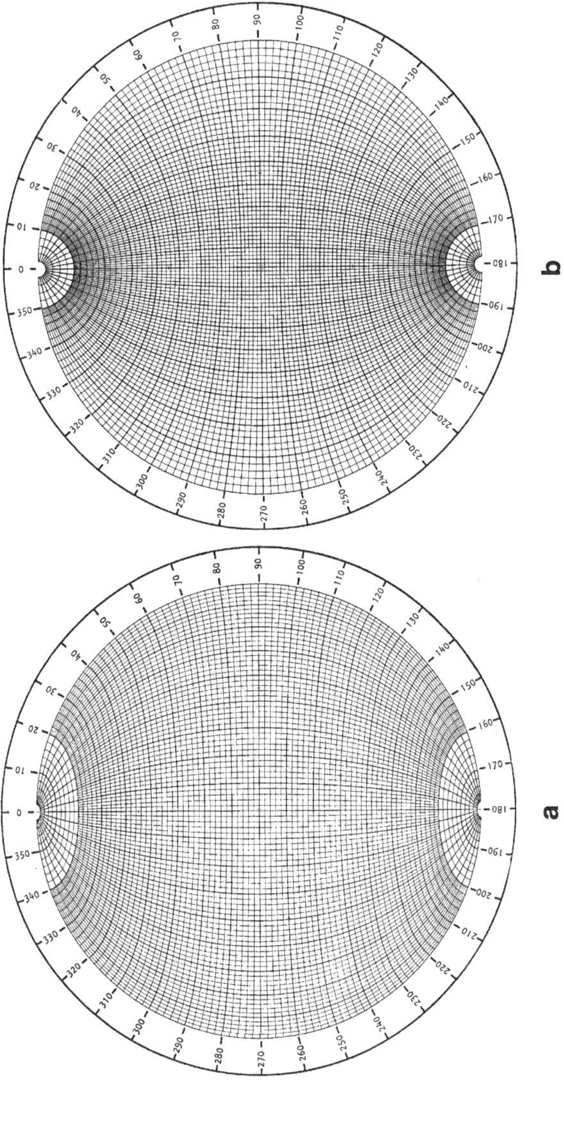

Fig. 1.16 (**a**) Equal areal network after Schmidt and (**b**) equal angle network after Wulff

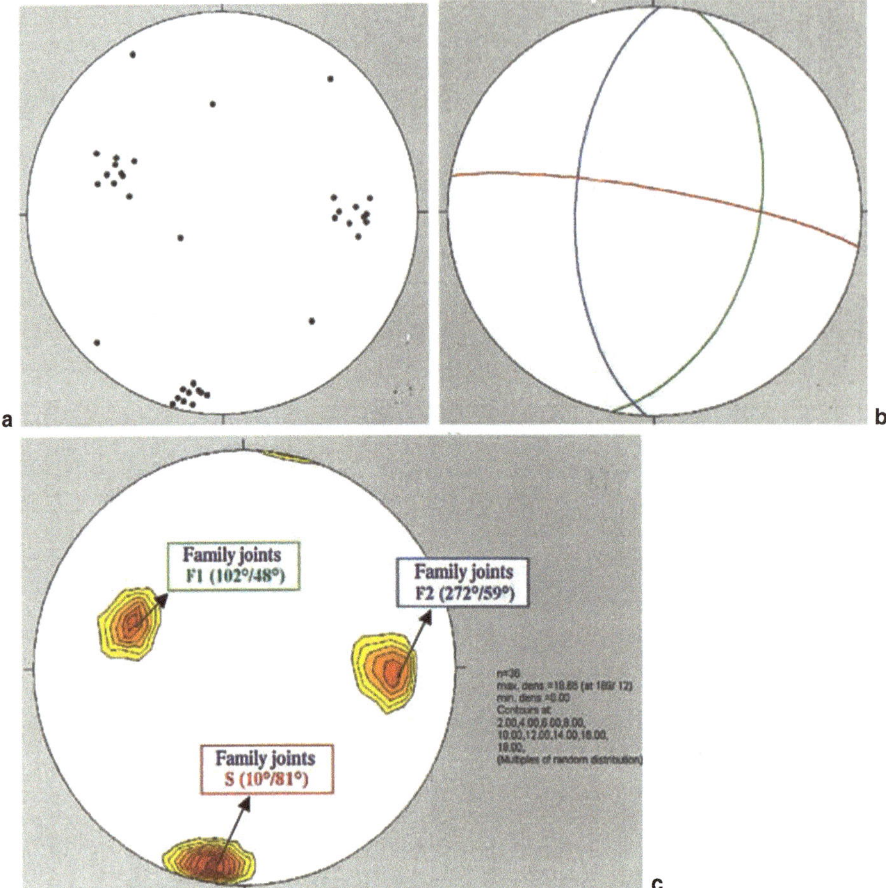

Fig. 1.17 Equal areal representation of the poles of discontinuities detected in a rock mass (**a**) and their contouring of equal pole density zones (**c**); the maximum pole concentrations of those zones represent the poles of the mean planes of families whose large circles are shown in the figure (**b**)

1.4.2 Equal Angle Projections

Wulff's projections keep constant the angles between discontinuities (see Fig. 1.16b).

With this type of representation it is possible to know the shapes and volumes of blocks that are formed combining the different discontinuity families. Wulff's projections are actually conceived in a way that makes it possible to visualize and calculate angle ratio with precision.

In Figs. 1.18 and 1.19 some exemplifying cases are presented about the determination of the geometry and unitary volume of blocks; the geometric figures that are generated can be regular, as in the example, or irregular.

Example 1 Geological-structural survey in situ

Rock outcrop on which geological-structural survey was made.

Beyond the stratification, three discontinuity families were detected (F_1, F_2 and F_3). For each one, following elements were measured: orientation, spacing, aperture, persistence, shape, weathering, roughness, filling, intercept, presence of water. Mean values ($\bar{X}$) and mean quadratic deviation of orientation, aperture and spacing were calculated. Then, discontinuities were represented (both as poles and great circles) on a stereographic projection.

Lithology: "Dolomia Principale"

Stratification			F_1			F_2			F_3		
Orientation (°)	Spacing (cm)	Aperture (m)	Orientation (°)	Spacing (cm)	Aperture (m)	Orientation (°)	Spacing (cm)	Aperture (m)	Orientation (°)	Spacing (cm)	Aperture (m)
288/50	18	0.0001	192/68	28	tight	50/70	4	0.0003	72/70	5	0.0002
295/50	10	0.00002	180/72	8	0.00001	52/78	4	0.0001	72/66	1	tight
288/50	10	0.0002	188/72	3	0.001	42/80	3	0.00005	78/82	1	0.001
306/60	15	0.0001	206/50	14	0.0002	56/90	2	0.0003	74/48	1	0.00004
300/60	8	0.00001	198/62	5	tight	30/70	5	0.0003	72/62	2	0.00003
284/65	8	0.0002	190/90	8	0.00003	44/80	2	0.0001	78/40	3	0.0001
290/60	6	0.0001	182/80	6	0.0002	48/78	3	0.002	68/58	4.5	0.0002
288/50	14	0.00003	190/75	22	0.0001	50/70	4	0.001	68/44	2.5	0.0003
290/48	17	0.00003	186/72	3	0.001	40/78	4	0.0002			
		208/58	2	tight	40/86	3	0.00001				
		180/70	1	0.0002	50/82	3	0.0003				
		180/84	7	0.0001	42/86	3	0.0002				
		184/82	2	0.0002	50/58	2	0.0001				
		184/80	3	0.0001							
$\bar{X}$ =	$\bar{X}$ =	$\bar{X}$ =	$\bar{X}$ =	$\bar{X}$ =	$\bar{X}$ =	$\bar{X}$ =	$\bar{X}$ =	$\bar{X}$ =	$\bar{X}$ =	$\bar{X}$ =	$\bar{X}$ =
292/55	11.77	$8.7\ 10^{-4}$	189/73	8	$2.2\ 10^{-4}$	46/77	3.32	$3.5\ 10^{-4}$	73/59	1.66	$2.3\ 10^{-4}$
σ =	σ =	σ =	σ =	σ =	σ =	σ =	σ =	σ =	σ =	σ =	σ =
6.9/6.3	4.32	$7.3\ 10^{-4}$	9.2/11	8.04	$3.4\ 10^{-4}$	6.9/8.6	0.92	$5.4\ 10^{-4}$	3.8/14.2	1.64	$3.2\ 10^{-4}$

Stratification	F_1	F_2	F_3
Persistence	**Persistence**	**Persistence**	**Persistence**
<50	50<PL<90	50<PL<90	<50
Shape	**Shape**	**Shape**	**Shape**
ondulating	stepped	stepped	stepped
Weathering	**Weathering**	**Weathering**	**Weathering**
weathered	weathered	weathered	weathered
Filling	**Filling**	**Filling**	**Filling**
granular	–	–	–
Water	**Water**	**Water**	**Water**
wet	dry	dry	dry
Vertical intercept	**Horizontal intercept**	**URV min - URV max**	**URV shape**
5.26 cm	2.38 cm	$4\ cm^3 – 750\ cm^3$	parallelepiped
Roughness (JRC)	**Roughness (JRC)**	**Roughness (JRC)**	**Roughness (JRC)**
13	15	9	5

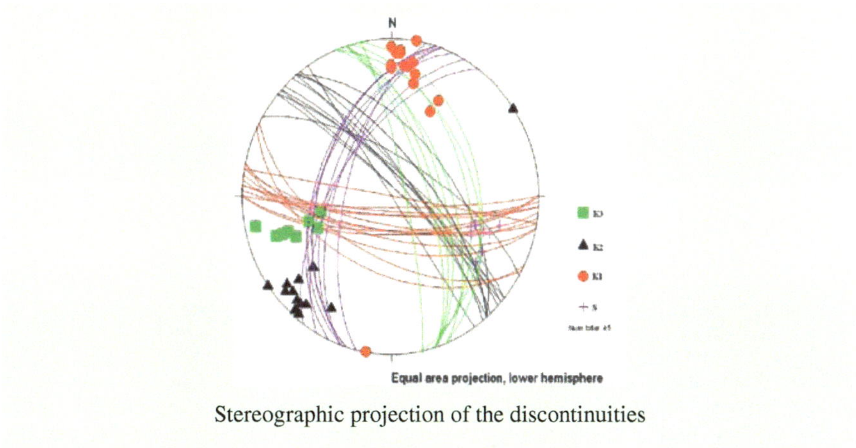

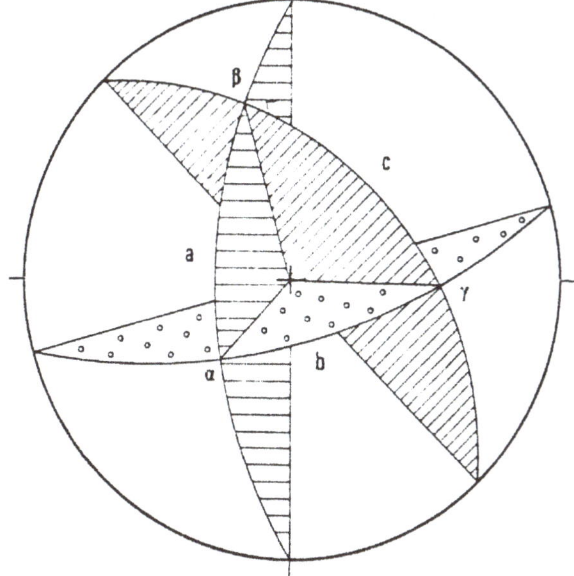

Fig. 1.18 Equal angular stereographic representation of three secant planes having following orientation: Plane (**a**) 270°/60°; plane (**b**) 164°/66° and plane (**c**) 46°/40°

1.5 Basic Elements for Hydrogeological Conceptual Model Definition

The definition of the hydrogeological conceptual model in a rock mass has to consider the following aspects:

- the work scale;
- the fracturing degree (e.g. the elementary representative volume);
- the changing of fracturing degree in the space, in particular with depth.

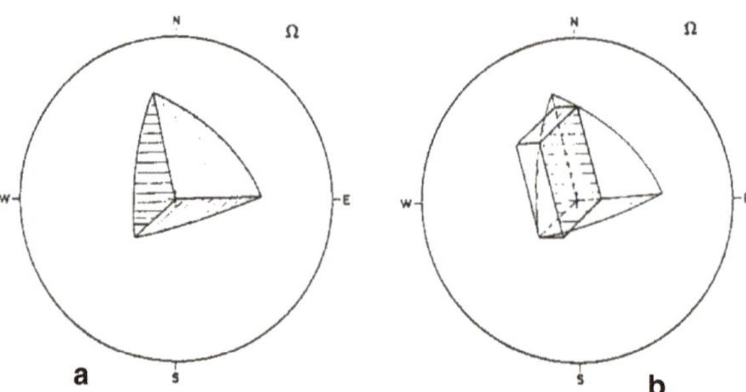

Fig. 1.19 Equal angular representation of the unitary block on the stereogram. It shows the following: (**a**) The polyhedron isolated by the three discontinuity families and the angles they enclose and (**b**) the shape and unitary volume of the block, determined according to the discontinuities spacing

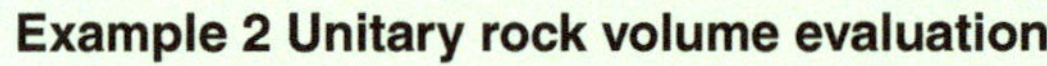

Example 2 Unitary rock volume evaluation

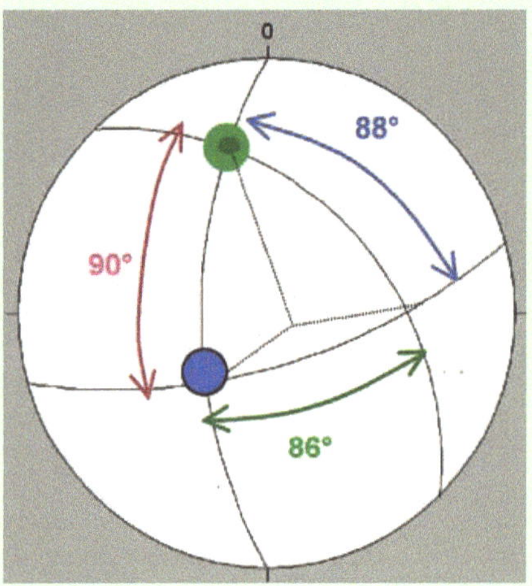

F_1: 270° / 60°, $S_1 = 0.5\text{m}$
F_2: 164°/66°, $S_2 = 1.2\text{m}$ S_i = spacing
F_3: 46° / 40°, $S_3 = 4.0\text{m}$

$$URV = \frac{S_1 S_2 S_3}{sen\alpha\, sen\beta\, sen\gamma} \qquad URV = \frac{0.5 \cdot 1.2 \cdot 4}{sen(90) \cdot sen(86) \cdot sen(88)} = 2.41\ \text{m}^3$$

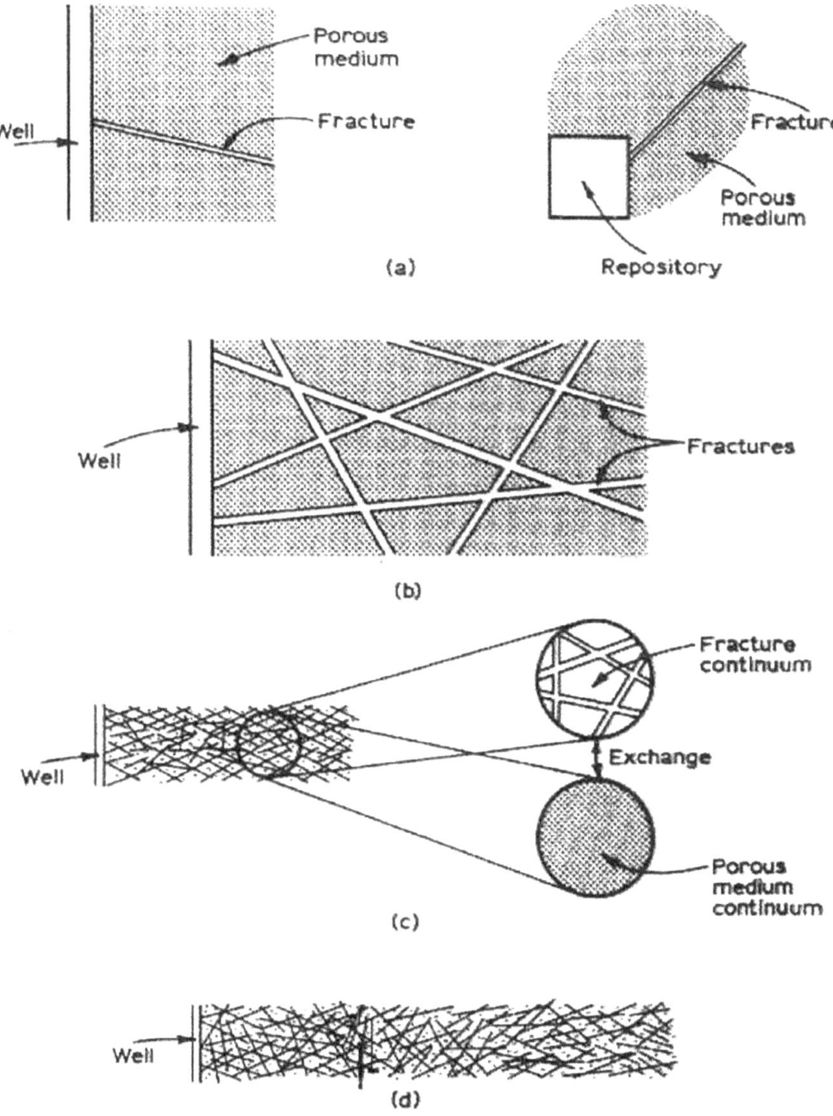

Fig. 1.20 Schematic representation of the different flow scale in a fractured medium: (**a**) Very near field; (**b**) near field; (**c**) far field; and (**d**) very far field (Bear and Berkowitz, 1987)

1.5.1 The Work Scale

The problem of percolation in a fractured medium is faced with different approaches according to the work scale (Fig. 1.20):

- *very near field*: The water flow occurs inside a single fracture;

- *near field*: The water flow occurs inside a small number of well-defined fractures, whose shapes and position are known, and that may also be reconstructed with a statistical approach;
- *far field*: The water flow occurs in two superposed continua, the first one is given by the system of fractures and the second one by porous blocks;
- *very far field*: The water flow occurs inside a fractured porous medium that can be assimilated to a continuum.

Evidently, the first two work scales are referred to well-localized problems. The second scale is often used; it requires detailed information on all discontinuities in the domain; as an alternative, if the statistical distributions relative to the fracture features (aperture, length, orientation, spacing etc.) are known, the probabilistic generation of discontinuity networks with Monte Carlo simulation methods (Min et al., 2004) could take place.

For problems on lower scales, continuous approaches are required; the choice depends not only on the work scale, but also on the fracturing features of the rock mass (La Pointe et al., 1996) that define the "elementary representative volume" (ERV).

1.5.2 Elementary Representative Volume

The "elementary representative volume" ERV is the minimal volume to take into consideration to study the hydrogeological features of a rock mass, such that the medium can be considered sufficiently homogeneous and isotropic (Bear, 1972).

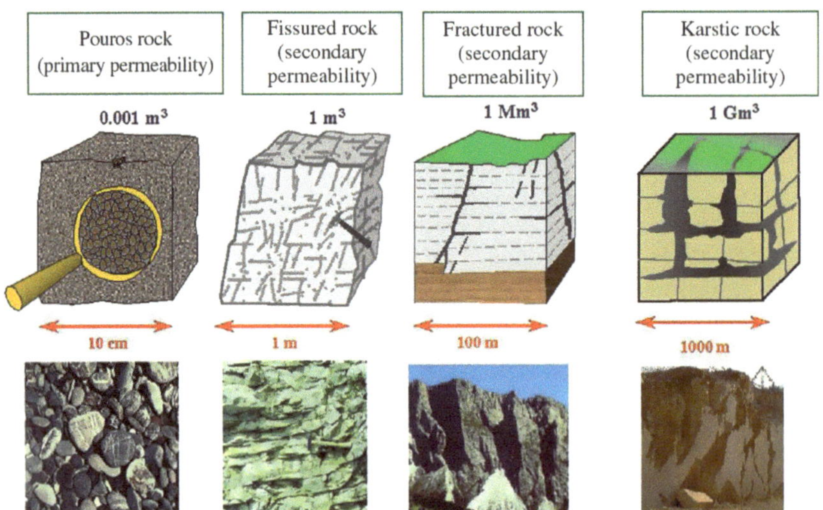

Fig. 1.21 Comparison among the elementary representative volumes (ERV) of different types of rocks (Civita, 2005)

In a rock mass, the ERV must contain all the different discontinuity families and therefore it can change, according to the fracturing degree, from 1 m^3 to $10^6 m^3$ (Fig. 1.21).

It is evident that the ERV dimension is strictly correlated to the URV and to the volumetric frequency of joints (Palmstron, 1982):

$$J_v = \frac{1}{s_1} + \frac{1}{s_2} + ... + \frac{1}{s_n}.$$

That relation is based on the knowledge of average spacing, s_i, of the different discontinuity families and on the number of families n.

The traditional approach is often difficult to apply due to the relevant variability of fracturing degree of rocks, which is generally studied using a statistical approach that makes use of fractals to define the discontinuity network (Mojitabai et al., 1989; Poulton et al., 1990; Ghosh, 1990). The scale invariance typical of fractals implies the existence of a law binding the number and dimensions of discontinuities:

$$N(r) \infty r^{-D}.$$

where r is the length of discontinuities, $N(r)$ the number of discontinuities having length over r and D is the fractal dimension that generally is variable for rocks from 1.14 to 3.54 (Turcotte, 1986) and, according to some authors, depends on the mechanical characteristics of the rock mass (Catani, 1999).

1.5.3 Changing of Fracturing Degree with Depth

Experimental observations show a reduction in the frequency of fracturing as depth increases, linked with a reduction of the aperture of joints (Snow, 1969). This determines an often quite relevant decline of the effective fracture porosity with depth, with a correlated decline in the permeability of the medium.

In reality, the definition of the trends in depth of the above mentioned geometric features of discontinuities is often quite difficult and experimental tests carried out by different authors during the years have often provided contrasting results. According to some authors, for example, beyond a superficial unit of weathered and ruined rock, where the fracturing frequency is higher, permeability remains almost constant in depth (Raven and Gale, 1976; Kendorski and Mahtab, 1976; Manev and Avramova-Tacheva, 1970); permeability changes should therefore be linked more to aperture changes than to variations in the intensity of fracturing, even though the presence of wide open joints in depth is not uncommon, with extremely variable permeability values that could not be easily reproduced by means of empirical laws, in particular, when depth exceeds 500 m (Brace, 1980).

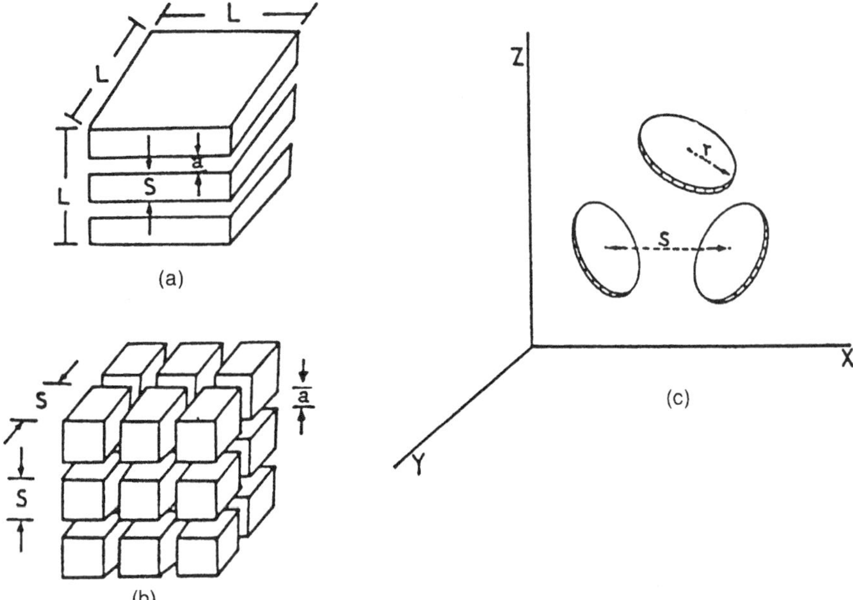

Fig. 1.22 Representation of discontinuities: (**a**) Model with a single discontinuity system; (**b**) according to orthogonal model; and (**c**) according to Baecher's theory (Lee and Farmer, 1993)

1.6 Probabilistic Generation of Discontinuity Network

Geometric characteristics of discontinuity systems can be represented through stochastic models as each parameter (orientation, spacing, etc.) can be described through statistical distributions that can be applied to statistically homogeneous parts of the rock mass (Aler et al., 1996; Rabinovitch et al., 1999).

All *stochastic models* share the following exemplifying hypothesis:

- all fractures are planar;
- for a fracture, all positions within the domain are equally probable;
- the orientation of each fracture is independent from its position.

Among the different models, the orthogonal model, Baecher Disks, Veneziano's model (1978) and Dershowitz's model (1984) must be remembered. Each one of these models introduces specific relationships between the different geometric characteristics of discontinuities, thus allowing to forecast and represent the geometry of the rock mass.

The *orthogonal model* (Imray, 1955) assumes the presence of two or more discontinuity systems, placed orthogonally to each other (Fig. 1.22b); spacing is considered constant or is assimilated to a Poissonian process, whereas the discontinuities can be of two types:

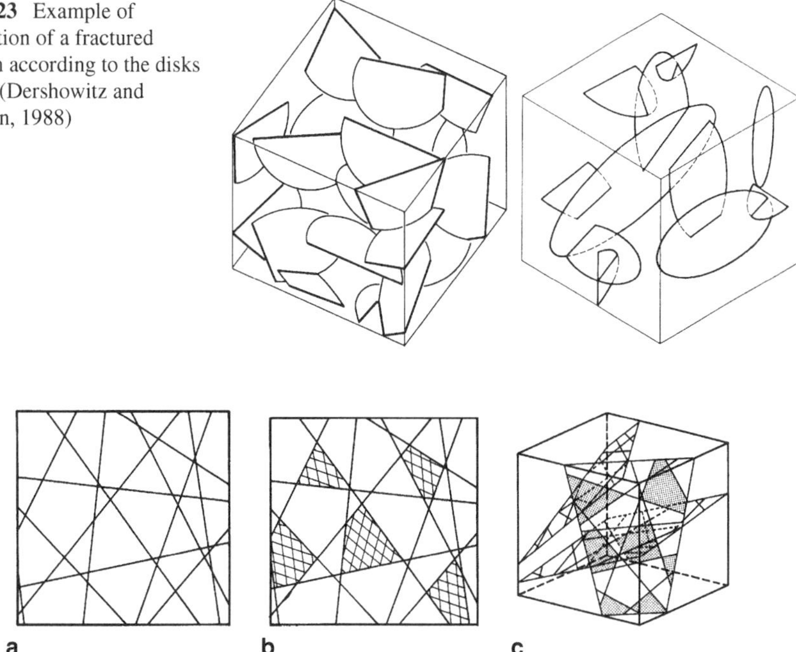

Fig. 1.23 Example of generation of a fractured domain according to the disks theory (Dershowitz and Einstein, 1988)

Fig. 1.24 Generation of a fractured model after Veneziano's (Dershowitz and Einstein, 1988). (**a**) Generation of traces on a plane using a poissonian process; (**b**) selection process of polygons and discontinuities; and (**c**) poissonian generation process of the fracture surface in 3-D

- unlimited (infinite length) that is with dimension over those of the considered domain;
- limited (with finite length), for which form and dimensions must be defined.

Baecher's theory (1977) assumes that the joints are round or elliptical (Fig. 1.22c); a discontinuity system is therefore originated by a number of parallel disks, whose centres have a random poissonian distribution in space, whereas the radius presents a lognormal distribution; the orientation of the disks can be considered constant or assimilated to an aleatory variable (Fig. 1.23); that theory was adopted and extended by many authors (Long, 1983; Billaux, 1990; Villaescusa, 1993).

Veneziano's model (1978) considers a poissonian distribution of the planes containing joints, for which a uniform distribution in space is supposed; a poissonian process is then used to divide the plane in regions having polygonal shape, which can be constituted by intact rock or contain joints (Fig. 1.24); the resulting traces of discontinuity present an exponential distribution.

Dershowitz's model (1984) is similar to the previous one concerning the generation of the joints planes; the intersections between the different planes define on each plane some polygons, which can be defined as joints or intact rock.

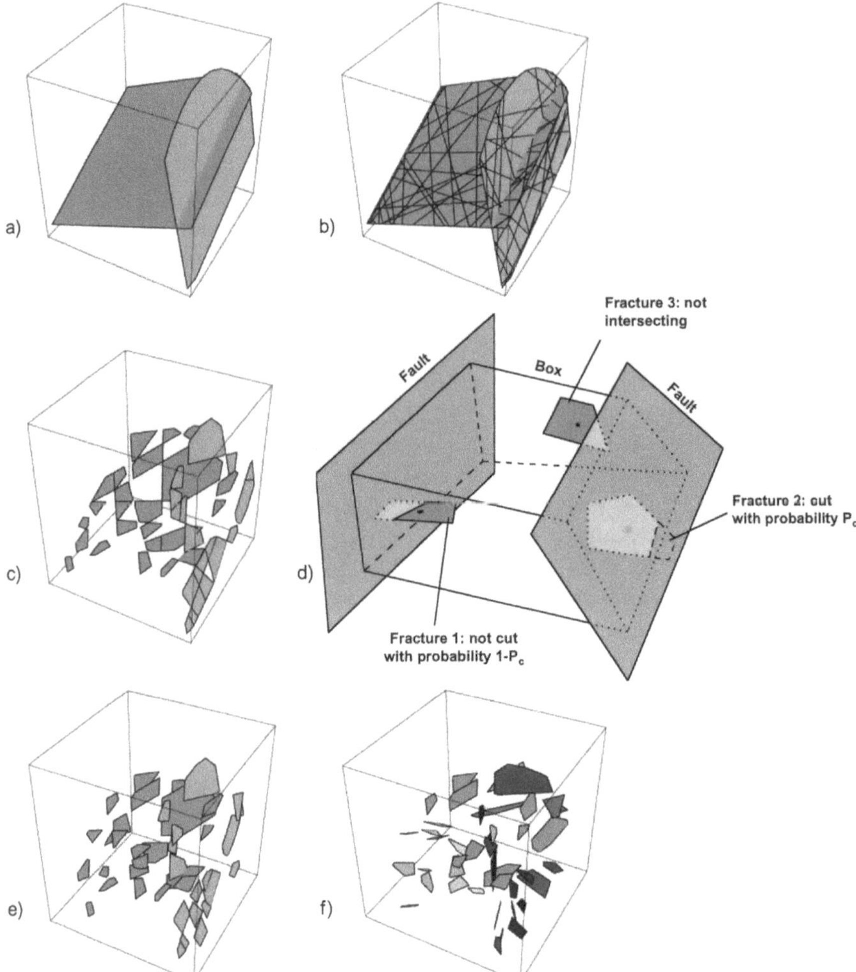

Fig. 1.25 Phases of the stochastic process to generate the geological-stochastic model (Meyer and Einstein, 2002). (**a**) **Phase 1**: It simulates the orientation of potential fracture surface using a poissonian model; (**b**) and (**c**) **phase 2**: Blocks of intact rock and discontinuities are detected on the fracture surfaces; (**d**) **phase 3**: It defines the different zone of the modellized volume, combining each one with a different intensity; as a consequence, it selects part of the polygons previously generated along the fracture surfaces; and (**e**) and (**f**) **phase 4**: Polygons can be transferred and/or rotated

Example 3 Discontinuity network generation

Based on the data obtained from the geological-structural in situ survey (Example 1), a number of different equally probable discontinuity networks were made using Dershowitz's model.

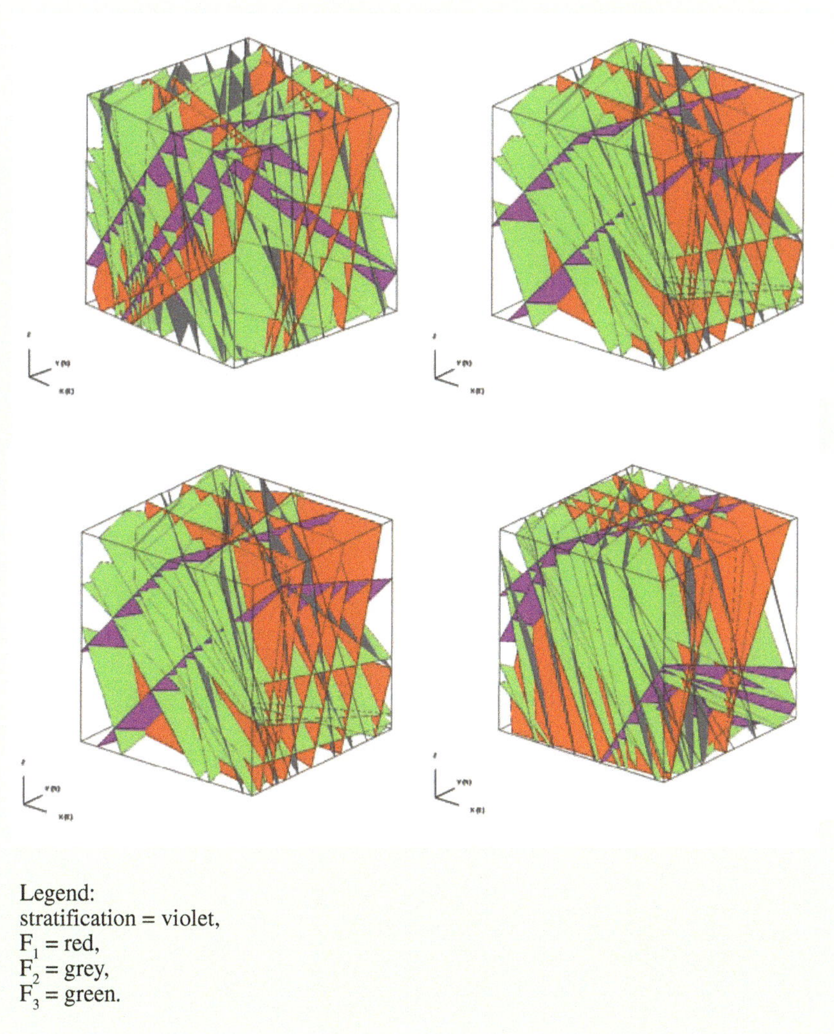

Following studies led to the implementation of stochastic models that could reproduce the fracture network using a stochastic process (Fig. 1.25), considering the main geological-structural features (Ivanova, 1998; Meyer and Einstein, 2002).

Chapter 2
Hydraulic Conductivity Assessment

2.1 Introduction

From the hydraulic point of view, rock masses are heterogeneous, anisotropous and discontinuous media. As water flow in rocks occurs mainly along discontinuities, the exact knowledge of their distribution and their characteristic parameters is fundamental to finding the features that describe the fluid flow, in particular, as far as the hydraulic conductivity assessment is concerned (that is the capability of a rock mass to be run through by water; Fig.2.1).

During the years, different deterministic and probabilistic methodologies have been developed to correlate structural data and hydrogeological parameters.

2.2 Deterministic Methodologies

2.2.1 Hydraulic Conductivity Along a Single Fracture

According to the deterministic methodologies, rock is considered permeable only through the fractures. From a hydraulic point of view, this corresponds to assimilating the water flow in the joint to what occurs between two parallel or sub-parallel planes (Fig. 2.2). Actually, starting from the indefinite Navier-Stockes equation

$$\rho\left(F - \frac{du}{dt}\right) = gradp - \eta\Delta_2 u,$$

where

- $F =$ the force due to the gravitational field;
- $p =$ pressure of the fluid;
- ρ and $\eta =$ density and dynamic viscosity of the fluid, respectively, and $u =$ its speed,
  assuming a uniform regime, following quantity can be obtained:
- piezometric gradient: $J = -\frac{d}{dx}\left(z + \frac{p}{\gamma}\right) = -\frac{\eta}{\gamma}\frac{d^2u}{dy^2}$;

L. Scesi, P. Gattinoni, *Water Circulation in Rocks*, DOI 10.1007/978-90-481-2417-6_2,

Conductivity, m/s.	10^{-1}	10^{-2}	10^{-3}	10^{-4}	10^{-5}	10^{-6}	10^{-7}	10^{-8}	10^{-9}
Degree of Conductivity	V. High	High		Moderate			Low		V. Low
Soil Type	Gravel		Sands		V. Fine Sands, Silts, Glacial Tills, Stratified Clays			Homogeneous Clays	
Rock Type	Shale Fractured — Sandstone Soln. Cavities — Limestone • Dolomite — Unfractured Cavernous / Fractured — Basalt — Dense Fractured/Weathered — Volcanics excl. Basalt Weathered — Metamorphics Bedded Salt Weathered — Granitic Rocks								

Fig. 2.1 Representative hydraulic conductivity values for different rock masses (Isherwood, 1979)

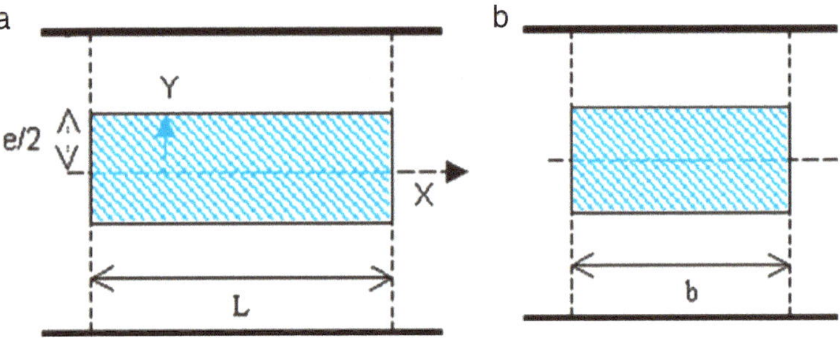

Fig. 2.2 Representation of the fluid flow between two parallel, indefinite and smooth planes, with constant aperture "e". Longitudinal section (**a**) and transversal section (**b**) (from Citrini and Noseda, 1987, modified)

- max speed along the middle plane: $u_{\max} = \frac{\gamma \cdot J}{8\eta} e^3$;
- fluid average speed: $V = \frac{Q}{be} = \frac{\gamma \cdot J}{12\eta} e^2$;
- flow rate of the whole section: $Q = 2 \int_0^{e/2} ubdy = \frac{\gamma \cdot Jb}{\eta} \int_0^{e/2} \left(\frac{e^2}{4} - y^2 \right) dy = \frac{\gamma \cdot J}{12\eta} be^3$,

where

- γ is the fluid specific weight;
- e and b are the aperture and the width of the discontinuity plane, respectively (Fig. 2.2).

Hence it follows that the hydraulic conductivity value, referring to a single smooth joint and in laminar conditions, is expressed as follows:

$$K = \frac{\gamma e^2}{12\eta} = \frac{g e^2}{12\nu},$$

Table 2.1 Relations used to calculate the hydraulic conductivity for different motion regimes.

Condition	Flor law	Unitary flow (m^2/2)	
Laminar	$\lambda = \frac{96}{R_e}$ (Poiseuille, 1839)	$q_i = \frac{g e_i^3}{12\nu} J_i$ (snow,1969)	(1)
Turbulent (smoot walls)	$\lambda = 0.316 R_e^{-1/4}$ (Blasius, 1913)	$q_i = \left[\frac{g}{0.079} \left(\frac{2}{\nu} \right)^{1/4} e_i^3 J_i \right]^{4/7}$	(2)
Turbulent (rough walls)	$\frac{1}{\sqrt{\lambda}} = -2 \log \frac{\frac{\varepsilon}{D_h}}{3.7}$ (Nikuradse, 1930)	$q_i = 4\sqrt{g} \left[\log \frac{3.7}{\frac{\varepsilon}{D_h}} \right] e_i^{1.5} \sqrt{J_i}$	(3)
Inertial	$\lambda = \frac{96}{R_e} \left[1 + 8.8 \left(\frac{\varepsilon}{D_h} \right)^{1,5} \right]$ (Louis, 1967)	$q_i = \frac{g e_i^3}{12\nu \left(1 + 8.8 \left(\frac{\varepsilon}{D_h} \right)^{1,5} \right)} . J_i$	(4)
Turbulent	$\frac{1}{\sqrt{\lambda}} = -2 \log \frac{\frac{\varepsilon}{D_h}}{1.9}$ (Louis, 1967)	$q_i = 4\sqrt{g} \left[\log \left(\frac{1.9}{\varepsilon / D_h} \right) \right] e_i^{1.5} \sqrt{J_i}$	(5)

R_e = Reynolds Number; ε = asperities height; $D_h = 2e$

where e is the joint mean aperture, g is the gravitational acceleration and ν, γ, η are the kinematic viscosity, the specific weight and the dynamic viscosity of the fluid, respectively.

If the *motion regime* (laminar, turbulent, inertial) and the joints' roughness in terms of relative *asperities height* are also taken into consideration, the relationships to determine the hydraulic conductivity are presented in Table 2.1 (Figs. 2.3 and 2.4). As can be observed, hydraulic conductivity strongly depends on the *joint's aperture* (e) and its *roughness* (ε / D_h).

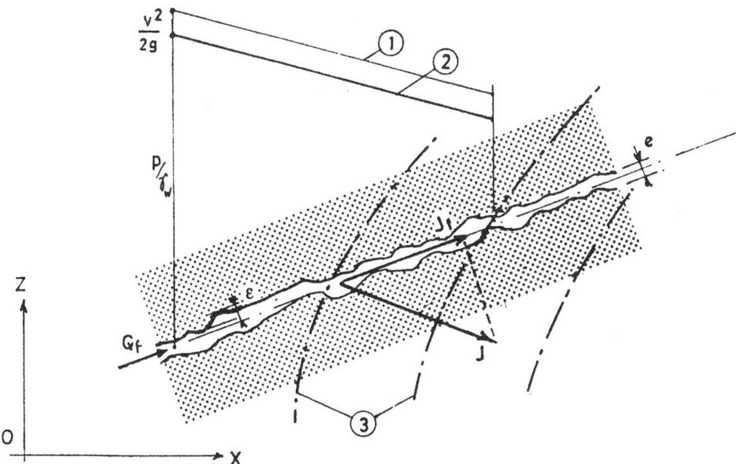

Fig. 2.3 Basic parameters for water flow characterization inside one fracture: Q_f is the unitary flow through the joint; e the joint aperture; J and J_f the hydraulic gradient and its component along the joint, respectively and ε the height of joint's asperities (modified from Louis, 1974)

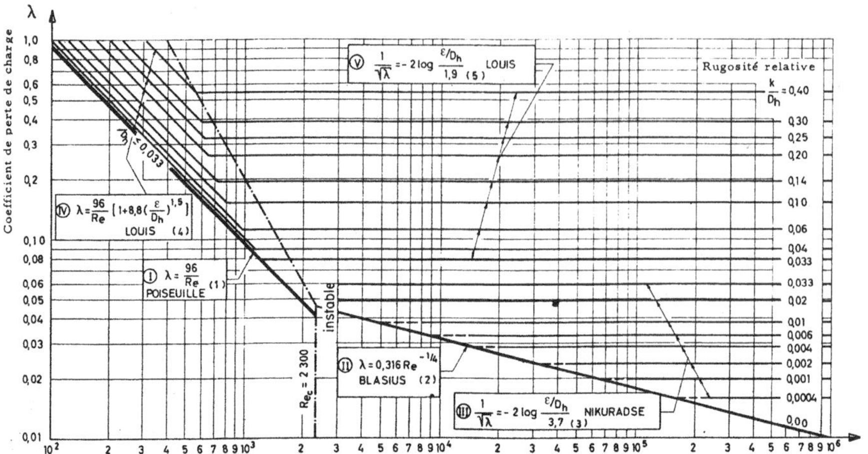

Fig. 2.4 Hydraulic laws for flow inside fractures (Louis, 1974)

The discontinuities apertures are generally small (≤ 1 mm), so that sometimes the asperities height is equal to the aperture width. Therefore, it is possible to distinguish between the following:

- a *superficial roughness* ($\varepsilon/D_h < 0.033$, according to Louis, the flow is parallel). The laws used for pipes are also valid in this case and the roughness influences just the turbulent regime;
- a *shape roughness* ($\varepsilon/D_h > 0.033$, according to Louis, the flow is not parallel). In this case, an inertial regime is generated and the convective accelerations transferred by the contouring shapes have to be considered, as they cause an average increment of the load loss along the whole discontinuity (Witherspoon et al., 1980), even though the linear term prevails over the quadratic one (*see* Table 2.1).

As regards the Reynolds' critical number (R_e), in rocks it assumes values far lesser than 2300 (De Marsily, 1986); some experimental tests (Louis, 1974) demonstrated that as ε/D_h increases, the critical R_e decreases and it verges on 100 if $\varepsilon/D_h \approx 1$. For many authors, R_e equal approximately 100 corresponds to the critical value for the filtration flows in porous media.

2.2.2 Hydraulic Conductivity Along a Fracture System

Discontinuities are never isolated, but they can be clustered in families; to consider also *spacing* and *frequency*, Snow (1969) proposed the following relation, which is true for smooth joints and laminar condition:

$$K_i = \frac{e_i^3 f_i g}{12\nu},$$

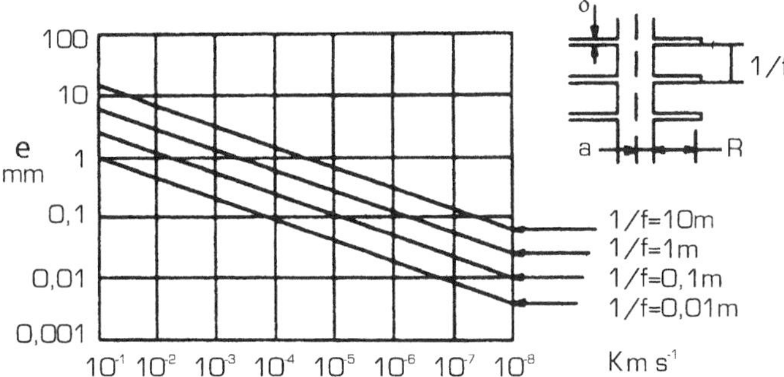

Fig. 2.5 Diagram to determine the hydraulic conductivity starting from aperture and spacing (Snow, 1969, modified)

Table 2.2 Evaluation of the aperture (in mm) of the fractures when spacing and hydraulic conductivity change (Francani, 1997)

$K(ms^{-1})$	1/f = 10 m	1/f = 1 m	1/f = 0.1 m	1/f = 10 mm
10^{-1}	11.5	5.3	2.5	1.15
10^{-2}	5.3	2.5	1.15	0.53
10^{-3}	2.5	1.15	0.53	0.25
10^{-4}	1.15	0.53	0.25	0.115
10^{-5}	0.53	0.25	0.115	0.053
10^{-6}	0.25	0.115	0.053	0.025
10^{-7}	0.115	0.053	0.025	0.011
10^{-8}	0.053	0.025	0.011	0.005

where f_i is the frequency (m^{-1}) of the ith discontinuity family (Fig. 2.5 and Table 2.2). Naturally, the single fracture hydraulic conductivity formula can be generalized and transformed in that for a discontinuity system; in a similar way, it is possible to obtain different formulas for all motion regimes in Table 2.1.

If joints with *filling* (sand, silt, clay, etc.) are present, hydraulic conductivity is conditioned by that material; the thicker the filling with respect to the joint's aperture, the stronger the influence. Granulometric and mineralogical analysis must be carried out to take this parameter into account.

2.2.3 Hydraulic Conductivity Tensor

The evaluation of the hydraulic conductivity of a rock mass characterized by one or more discontinuities systems must take into account also the *orientation* of the discontinuities that, as already mentioned, influences the water flow; therefore, hydraulic conductivity is expressed as a tensor (Louis, 1974; Kiraly et al., 1971):

$$\overline{\overline{K}} = \sum_{i=1}^{m} \left[\left(e_i^3 \cdot g \cdot f_i \right) \div 12\nu \right] \overline{\overline{A_i}},$$

where

- $\overline{\overline{K}}$= conductivity tensor (m/s);
- e = average discontinuities aperture (m);
- g= standard gravity acceleration (m/s^2);
- f = average frequency (number of discontinuities per length unit) (1/m);
- υ= water kinematic viscosity at 20°C (10^{-6}m^2/s);
- m= number of discontinuity families;
- $\overline{\overline{A}}$= $(\overline{\overline{I}} - \overline{n} \otimes \overline{n})$ orientation tensor, with $\overline{\overline{I}}$= fundamental tensor (identity matrix) and $\overline{n}$ = normal vector to the mean plane of the discontinuity family considered.

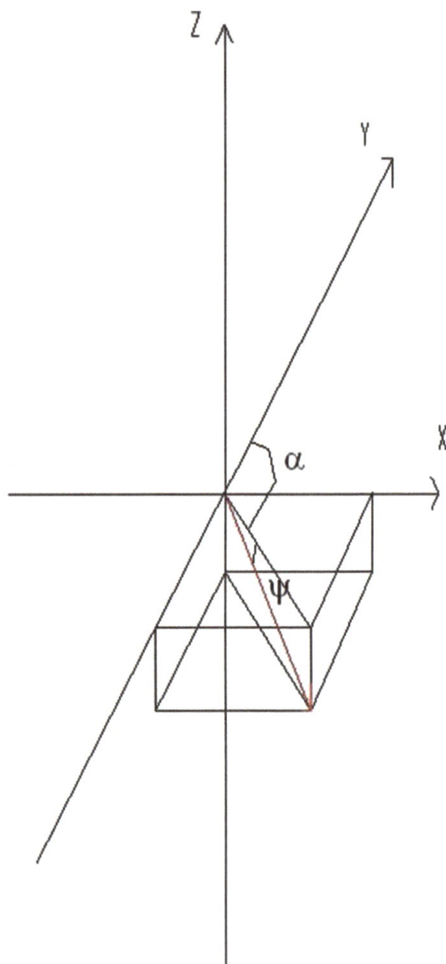

Fig. 2.6 The conventional reference frame XYZ for the hydraulic conductivity tensor orientation: α = dip direction (angle between the projection of the normal exiting from the plane on the horizontal plane and the North, calculated clockwise) and ψ = dip angle (angle between the discontinuity plane and the horizontal plane)

The vectorial product $(\bar{\bar{I}} - \bar{n} \otimes \bar{n})$ gives a second order tensor oriented in a preset reference system (Fig. 2.6), whereas the hydraulic conductivity projection in the reference system considered can be obtained from the product K[$(\bar{\bar{I}} - \bar{n} \otimes \bar{n})$]:

$$K \begin{vmatrix} (1-x^2) & -xy & -xz \\ -yx & (1-y^2) & -yz \\ -zx & -zy & (1-z^2) \end{vmatrix}.$$

Calculating the eigenvalues and the eigenvectors of the thus obtained matrix, the three main hydraulic conductivities, K_1, K_2, K_3 ,and their orientation in space are obtained (Fig. 2.7); with these results, it is then possible to build the conductivity ellipsoid having K_1, K_2, K_3 as semiaxes (Fig 2.8) and, therefore, the rock mass can be considered as an anisotropous continuum.

2.2.4 Equivalent Hydraulic Conductivity

If the relation (a geometric mean) proposed by Louis (1974) is used, hydraulic conductivities along the main strikes can be transformed into equivalent hydraulic conductivity (K_e):

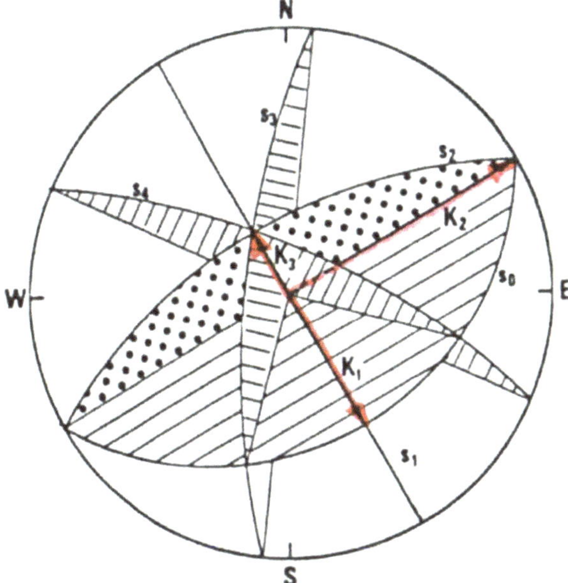

Fig. 2.7 Stereographic representation of discontinuity families (s_0, s_1, s_2, s_3 and s_4), characterized by constant aperture ($e_0 = e = e_2 = e_3 = e_4 = 0.1$ mm) and variable frequency ($f_0 = 4; f_1 = 2; f_2 = 1; f_3 = f_4 = 0.5$) and relevant main hydraulic conductivities, characterized by a module ($K_1 = 5.4 \times 10^{-6}$ m/s; $K_2 = 4.4 \times 10^{-6}$ m/s; $K_3 = 3.3 \times 10^{-6}$ m/s) and a strike in space

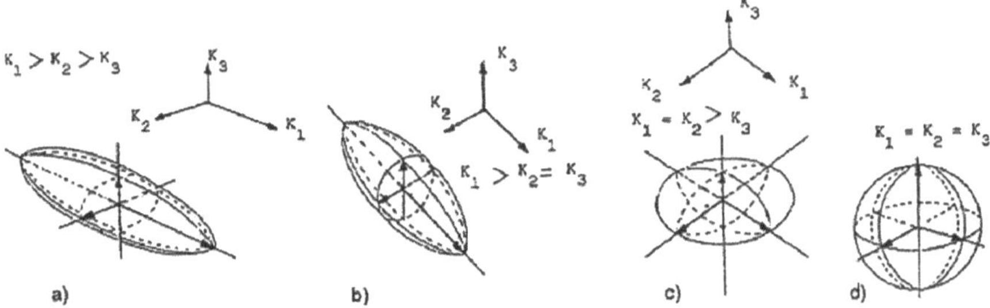

Fig. 2.8 Reference ellipsoid for the hydraulic conductivity tensor for the different anisotropy levels (**a, b, c**) and for an isotropous mass (**d**)

$$K_e = (K_1 K_2 K_3)^{1/3}$$

The value thus obtained can be used as representative value of the average hydraulic conductivity of the rock mass considered as an isotropous continum.

Using a stochastic approach to build the ERV and calculating the corresponding equivalent hydraulic conductivity, the real ERV dimension can be determined (Min et al., 2004).

2.3 Probabilistic Methodologies: Percolation Theory

Some authors (Shante and Kirkpatrick, 1971; Gueguen and Dienes, 1989) developed a so-called percolation theory to calculate the hydraulic conductivity starting from structural data. According to that theory, single discontinuities in rock masses are assimilated to bonds with finite length, whose hydraulic conductivity is a function of the aperture, whereas their intersections create the "percolation site". In this way, the real extension of the discontinuity inside the rock mass (*persistence*) can be considered. Discontinuities create a percolation network that, in the bidimensional space, is characterized by a number Z of bonds that leave each "site".

These bonds can be closed (the fluid cannot flow through a closed site and it can neither wet or pass through a blocked site), form a localized cluster or create a percolation path (Fig. 2.9).

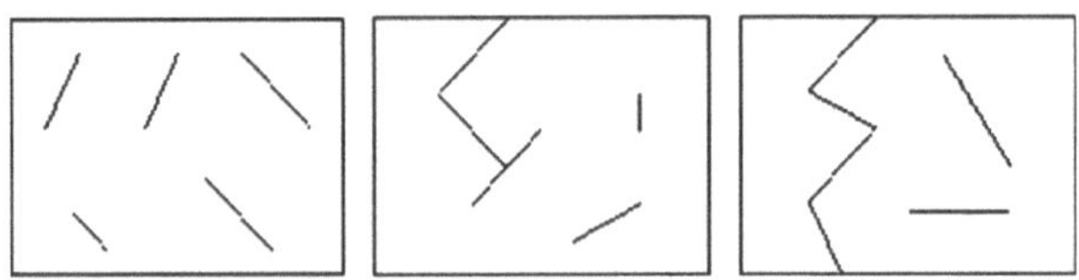

Fig. 2.9 Examples of percolation networks

Example 1 Determination of the hydraulic conductivity starting from data obtained in situ and provided in example 2 (Chapter 1)

Parameter	Stratification	F_1	F_2	F_3
Dip direction [°]	292	189	46	73
Dip [°]	55	73	77	59
Frequency [1/m]	8.4962	12.5	30.1205	60.2410
Aperture [m]	0.87×10^{-3}	2.2×10^{-4}	3.5×10^{-4}	2.3×10^{-4}
JRC	13	15	9	5
Regime	Laminar parallel	Turbulent (rough walls)	Laminar parallel	Laminar parallel
K_i [m/s]	4.57×10^{-3}	5.09×10^{-4}	1.06×10^{-3}	5.9×10^{-4}

The hydraulic conductivity tensor is equal to:
$K_1 = 6.6 \times 10^{-3}$ m/s (180°/0°)
$K_2 = 5.9 \times 10^{-3}$ m/s (277°/22°)
$K_3 = 1.0 \times 10^{-3}$ m/s (117°/66°)

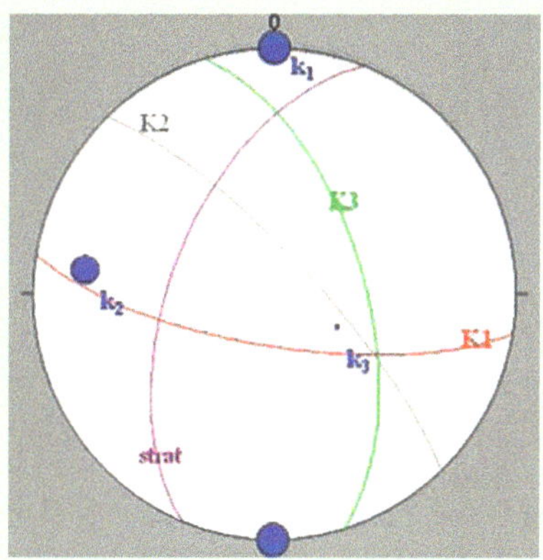

The corresponding equivalent hydraulic conductivity is equal to:
$K_{eq} = 3.39 \times 10^{-3}$ m/s

In this case, there is a probability, called $P_{(p)}$, that the fluid will flow through an infinite number of sites leaving from a source site at random:

$$P_{(p)} = \lim_{n \to \infty} P_n(p),$$

where $P_n(p)$ is the probability that a single source site will wet at least n other sites.

Beyond probability $P_{(p)}$, following probabilities can be defined as well:

- probability p, that is the probability that an arbitrary site being occupied by water;
- the critical probability p_c, which can be defined as

$$p_c = SUP\left\{p, such\ as\, P_{(p)} = 0\right\}.$$

According to these definitions, if the probability "p" is lesser than p_c, the water may spread only locally, i.e. all clusters of wet sites are finite in size; if, on the contrary, the probability "p" is higher than p_c, at least one percolation channel can spread through the medium.

Adopting a similar logical process, one can define a *percolation factor* (ξ) that is dependent on the dimensions (length) and/or density of fractures in the medium. That factor is expressed through the following empirical relation:

$$\xi = \pi^2\left(\frac{r}{s}\right)^3,$$

where

- s = average discontinuity spacing;
- r = average discontinuity radius (obtained from the average discontinuity length[1]).

[1] According to various Authors, the average length of discontinuities belonging to the same family can be calculated using the empirical formula (Phal, 1981):

$$L = \frac{l(h - 2\eta)(1 + \theta_r - \theta_c)}{(l\cos\phi + hsen\phi)(1 - \theta_r + \theta_c)}$$

where:

- l = length of the outcropping;
- h = height of the outcropping;
- η = minimum length considered in measuring discontinuity traces (*cut-off*);
- ϕ = complementary angle of the apparent average dip of the discontinuity family;
- $\theta_r = N_T/N$ e $\theta_c = N_c/N$;
- N_t= number of discontinuities going through the outcrop;
- N_c= number of discontinuities within the outcrop;
- N = number of discontinuities intercepting the outcrop.

The necessary data for the application of that relation can be easily obtained from a detailed analysis of pictures taken using digital cameras. Then, the model proposed by Baecher et al. (1977) is adopted as the representative model of the discontinuity traces – that is discontinuities are represented by circles or ellipses whose centres are placed in any point in space and whose radiuses present a lognormal distribution – the average discontinuity length is expressed as the average radius: $r = 2L/\pi$.

The percolation factor shows the *connectivity* degree, which is the interconnection among intersecting fractures that belong to different families; this interconnection allows a fluid particle to flow within the interconnected system moving from one fracture with finite length to the other.

Therefore, the *effective percolation probability* is a function of "connectivity" and of the "considered rock volume":

$$\nu = f\,(ERV, \xi)\,,$$

where

- ν = percolation frequency;
- ERV = elementary representative volume.

The *ERV* parameter introduces the scale effect; actually, an increase in the measurement scale will correspond to an increase in the percolation frequency. It is, therefore, necessary to consider the dimensions of the sampled volumes in connection to the fracturing degree.

In two-dimensional cases and for regular lattices (square, rectangular, honeycomb, etc.), in particular, assuming that 4 discontinuities ($z = 4$; Fig. 2.10), leave one site, the percolation frequency is given by

$$\nu \cong 54\,(p - p_c)^2$$

where

- p = probability that two fractures may intersect $\cong \pi^2 r^3/4s^3$;
- $p_c \cong 1/(z-1)$, with z = number of bonds leaving a single site.

ν may range from 0 and 1 (non-conductive and conductive, respectively) and it expresses the ratio between the percolation paths through which the flow occurs and the number of clusters. For highly fractured rock masses ($\xi > 10$), $\nu = 1$ will be set.

To calculate the permeability, expressed in m^2, the following relation will be used:

$$k = \frac{4\pi \nu e^3 r^2}{15 s^3}[\text{length}^2],$$

where

- ν = percolation frequency of an infinite network, determined from percolation theory;
- e = effective aperture;
- $r = 2L/\pi$ (with L equal to the average discontinuity length);
- s = average discontinuity spacing ($1/f$).

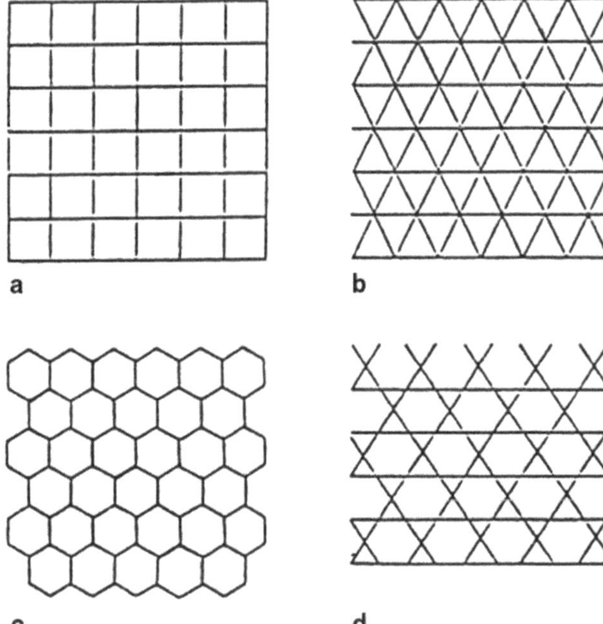

Lattice	Z	p_{cb}	Zp_{cb}	f_s	p_{cs}	$f_s p_{cs}$
Honeycomb	3	0.6527	1.96	0.61	0.700	0.427
Kagomé	4	—	—	0.68	0.653	0.444
Square	4	0.5000	2.00	0.79	0.590	0.466
Triangular	6	0.3473	2.08	0.91	0.500	0.455
Diamond	4	0.388	1.55	0.34	0.425	0.145
s.c.	6	0.247	1.48	0.52	0.307	0.160
b.c.c.	8	0.178	1.42	0.68	0.243	0.165
f.c.c.	12	0.119	1.43	0.74	0.195	0.144
h.c.p.	12	0.124	1.49	0.74	0.204	0.151

Fig. 2.10 Some common two-dimensional lattices and related critical probabilities and dimensional invariants: (**a**) square; (**b**) triangular; (**c**) honeycomb and (**d**) quinconce (Lee and Farmer, 1993)

Hydraulic conductivity can be obtained multiplying the permeability by the fluid density (ρ) and gravity acceleration (g) and dividing by the dynamic viscosity (η) of the fluid:

$$K = lg\rho/\eta[\text{length/time}].$$

2.4 In Situ Tests

2.4.1 Lugeon Tests

For the in situ determination of hydraulic conductivity in rock masses, "Lugeon" tests are usually carried out (Fig. 2.11).

They consist of pumping water under pressure in boreholes with a diameter ranging from 50 to 150 mm. A pipe to pump water is let down in the borehole; packers will allow the isolation of the borehole section L where permeability is being measured.

During the test, following parameters are measured: the injection pressure (with gauge), the inflow discharge and the time, after reaching the regime of permanent flow (10–20 minutes).

The test is carried out with a pressure loop characterized by 4÷5 loading pressure steps equal to 0.25 or 0.20 MPa, followed by a 3-step unloading phase. With due caution, that is to say, preventing water to flow again in the borehole, the effective pressure P_e (Pa) is given by the relation:

$$P_e = P_m + \gamma_w (H - H_p)$$

where P_m (Pa) represents the pressure indicated by the gauge, H (m) the height of the water column, H_p (m) the load loss in terms of water height and γ_w(N/m^3) the specific weight of water.

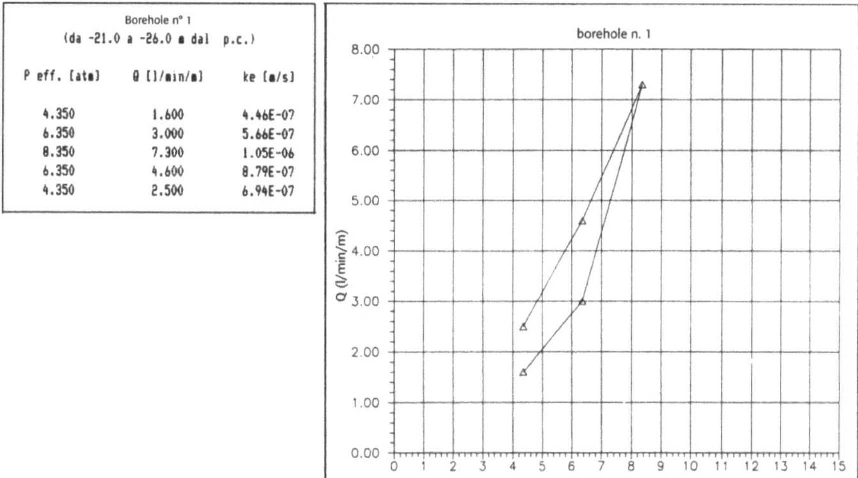

Borehole n° 1
(da -21.0 a -26.0 m dal p.c.)

P eff. [atm]	Q [l/min/m]	ke [m/s]
4.350	1.600	4.46E-07
6.350	3.000	5.66E-07
8.350	7.300	1.05E-06
6.350	4.600	8.79E-07
4.350	2.500	6.94E-07

Fig. 2.11 Example of Lugeon test. P_{eff} is the effective pressure and Q is the adsorbed flow rate

The load loss in the injection pipe must be calculated with relations apt for small diameter pipes, and, in case of very important tests, they will have to be determined experimentally or it will be necessary to measure the water pressure in the section being tested with appropriate equipments.

Much information on the fracturing state of the rock can be obtained by analyzing the morphology of the curve reconstructed by the Lugeon test.

As far as hydraulic conductivity measured in laminar flow condition is concerned, the following relation can be adopted:

$$K = \frac{\gamma_w Q}{P_e C_f},$$

where Q (m^3/s) represents the absorbed flow rate, P_e (Pa) the effective pressure and C_f (m) a shape coefficient given by the expression

$$C_f = 2 \cdot \pi \cdot D \cdot \frac{\sqrt{(L/D)^2 - 1}}{\ln\left(L/D + \sqrt{(L/D)^2 - 1}\right)},$$

where D and L are the diameter and the length, respectively, of the borehole section used for the test.

2.4.2 Hydrogeochemical Methods

Hydraulic conductivity can be determined in situ also using hydrogeochemical methodologies. By analyzing percolating water and groundwater, it is possible to draw some considerations about hydrochemical facies, exchanges with the aquifer matrix, residence and transit times, speeds of water flow, etc., from which the main hydrodynamic parameters can be obtained. Generally, three main hydrogeochemical methods are distinguished, which will be described later.

2.4.2.1 Traditional Geochemical Methods

Water circulating in rock masses interacts with the aquifer that contains it, thus modifying its chemical composition (Huizar et al., 1998; Jankowski et al., 1998). If the chemistry of feeding waters (rainfalls, superficial waters, seawater, irriguous water, etc.) is analyzed and confronted with that of groundwater, it is possible to determine its residence time (Edmunds and Smedley, 2000; Tweed et al., 2005). Actually, the different processes of hydrolysis, precipitation, salt dissolution, ionic absorbiment-exchange, oxidation, reduction, osmotic effect, etc. modify, more or less intensely, the chemistry of origin waters. Moreover, knowing the charge and discharge (springs) areas, through the residence time it is also possible to estimate the transit speed. If the piezometric gradient is known, these parameters allow to obtain the hydraulic conductivity in the medium.

It is evident that the traditional geochemical method implies a detailed knowledge of the hydrogeological setting of the examined area and the principles ruling the balance between rocks and watery solutions.

2.4.2.2 Methods with Artificial Tracers

The use of artificial tracers in the study of aquifer systems allows the identification of the paths followed by groundwater and, as a consequence, its circulation times and speeds. Among the most commonly used artificial tracers there are radioactive isotopes (trithio, chromium, bromine, iodine), chemical tracers (sodium chloride, bromides), biological and fluorescent tracers (fluorescein, rhodamine B, etc.). Usually, they are used in dosed solutions that are let in water flows, natural cavities, piezometers, water collector drains placed along underground excavations, etc. These solutions then spread in the aquifers and can be detected by specific isotope samplers placed at specific water points (springs, wells, outcrop waters, etc.) situated at lower levels in the valley than the input areas.

It is, therefore, possible to draw the underground paths followed by waters, the transit times and flow speeds. Moreover, analyzing the tracers response functions at a spring it is also possible to detect the different types of discontinuity and bond networks (Fig. 2.12).

The *unique impulse tracers response*, characterized by a curve with a main peak and a quick increase and decrease of the tracer in time, shows that the tracer follows one single path before reaching the spring, independently from the input point.

The *multiple impulse tracers response*, characterized by a sequence of peaks, reveals a quite complex bond network; actually, it takes the tracer quite a long time to reach the spring with different flow speeds and low concentrations, due to its dispersion in the discontinuity network.

The *dispersion impulse tracers response*, characterized by a curve with no peaks, reveals highly fractured rock masses and interconnected joints. The existence of real groundwater determines a relevant dilution of the tracer that reaches the discharge area with very low concentration and only after a very long time.

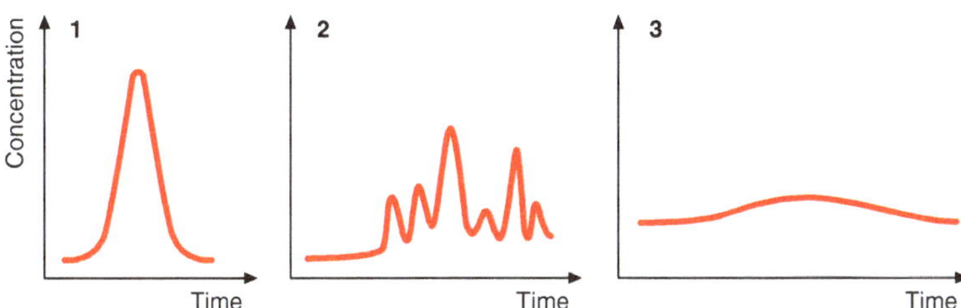

Fig. 2.12 Tracers response functions: (**1**) unique impulse; (**2**) multiple impulses; and (**3**) dispersion (Vigna, 2001)

2.4.2.3 Isotopic Methods

Trithio, a radioactive isotope of the water molecule with a decay time of 12–45 years, is often used to estimate the age of waterbodies characterized by a limited extension or by a quite fast flow. To use this "hydrogeological clock" in a proper way, the comparison between the trithio content of groundwater and that of rainfalls must be made. After the nuclear tests of the 1950s and the nuclear accident in Chernobyl in 1986, certain amounts of trithio, whose proportions vary in time, are present in the atmosphere, and as a consequence, in meteoric waters.

The distribution curve trithio content in rainfalls, according to the time, represent the "input" function of a hydric system. The comparison of the data of this curve with the changes of the activity in trithio in any water point, in time, allows to obtain the underground residence time of water. Then, the quantitative evaluation of the hydrogeological parameters is possible using modelistic approaches of mass transport.

2.4.3 Hydraulic Tests in Double-Porosity Aquifers

In some rock masses, e.g. carbonatic rock masses, two different "media", each one with its own physical peculiarities, may coexist: the intact matrix, characterized by primary porosity and low hydraulic conductivity, and the discontinuities, characterized by secondary porosity and high hydraulic conductivity (Barenblatt et al., 1960).

To determine the hydrogeological parameters in "double-porosity" aquifers the Kazemi method (1980) is generally used. The test is valid if the following conditions are present:

- the aquifer is confined and of almost infinite areal extent;
- the thickness of the aquifer is uniform over the area that will be influenced by the test;
- the well fully penetrates a fracture;
- the well is pumped at a constant rate;
- the piezometric surface is horizontal over the area that will be influenced by the test;
- a geometric schematization of the fracture system is possible (Fig. 2.13);
- each infinitesimal volume of the aquifer contains portions of both rock matrix and fractures;
- the rock matrix has a lower permeability and a higher storativity than the fracture system;
- the interporosity flow occurs in an almost steady state;
- the rock matrix and the fractures are compressible;
- $\lambda < 1.78$ (λ = interporosity flow coefficient).

The double-porosity theory states that the piezometric drawdown response to pumping can be expressed as follows:

$$s = \frac{Q}{4\pi T_f} F(u^*, \lambda, \omega),$$

where

$$u^* = \frac{T_f t}{(S_f + \beta S_m) \cdot r^2}, \qquad \lambda = \alpha \cdot r^2 \frac{K_m}{K_f}, \qquad \omega = \frac{S_f}{S_f + \beta S_m},$$

furthermore

- f = subscript indicating the fracture properties;
- m= subscript indicating the matrix blocks properties;
- T = transmissivity = $\sqrt{T_{f(x)} \cdot T_{f(y)}}$;
- S = storativity coefficient;
- K = hydraulic conductivity;
- t = time;
- r = well radius;
- λ = interporosity flow coefficient (dimensionless);
- α = shape factor, characteristic of the geometry of the fractures and aquifer matrix;
- β = geometric factor defining the schematization adopted ($\beta = 1/3$ for orthogonal system, $\beta = 1$ for strata type) (Fig. 2.13).

In the first moments of the tests, the equation expressing the water table drawdown response can be expressed as given below:

$$s = \frac{2.3Q}{4\pi T_f} \log \frac{2.25 T_f t}{S_f r^2}.$$

In the initial phases the water drawn toward the well only comes from the fracture system, therefore β is null, whereas in the final phases of the test the water also comes from the rock matrix and the drawdown response can be expressed as

$$s = \frac{2.3Q}{4\pi T_f} \log \frac{2.25 T_f t}{(S_f + \beta S_m) \cdot r^2}.$$

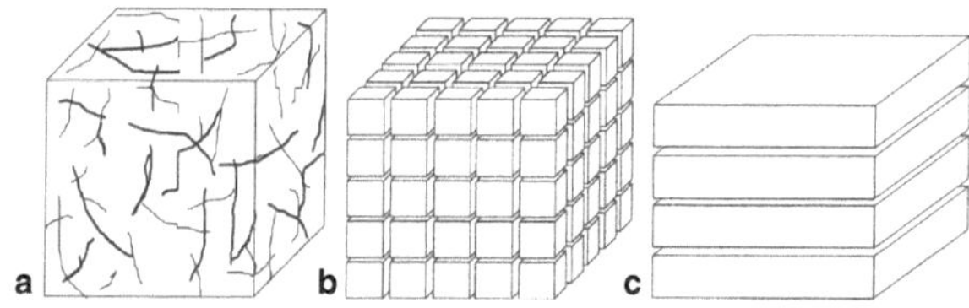

Fig. 2.13 Examples of fractured rock formations: (**a**) naturally fractured rock formation; (**b**) idealized three-dimensional, orthogonal fracture system; and (**c**) idealized horizontal fracture system

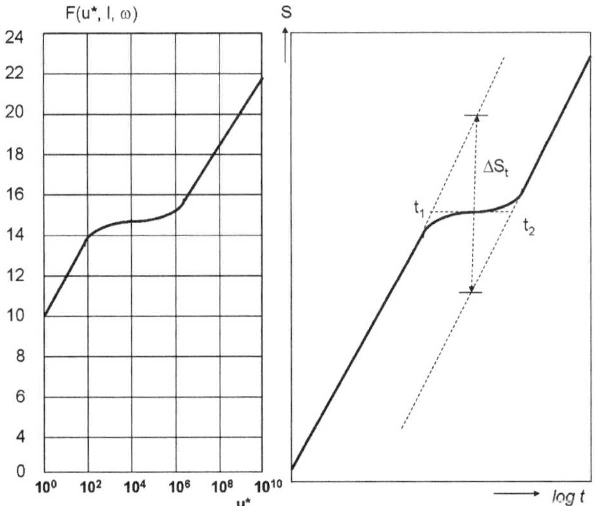

Fig. 2.14 (**a**) Theoretic curve representing the behaviour of a double-porosity aquifer; and (**b**) experimental curve

This expression is similar to Couper and Jacob's (1946), which only describes the drawdown behaviour in porous systems; actually, in the last pumping phases, it can be assumed that steady state set and that the aquifer response might be assimilated in a homogeneous and unconsolidated isotropic medium having transmissivity equal to that of the fractured system and a storativity coefficient equal to the arithmetic sum of the storativity coefficients of the fracture system and of the intact matrix (D'Aquino, 2006). The behaviour of double-porosity aquifers can be represented by the theoretic curve in Fig. 2.14a. Figures 2.14b and 2.15 represent the experimental curves used, together with the above-written expressions, to obtain the hydrogeological parameters required (for in-depth studies see the specific books in the Bibliography).

2.4.4 Hydraulic Tests in Anisotropic Aquifers

The analyses on hydraulic tests in anisotropous media (both on the horizontal and on the vertical plane) and in confined or unconfined aquifers are usually carried out using Papadopulos (1965), Hantush (1966), Neuman et al. (1984) and Weeks (1969) methods. Hantush's method, for example, allows the evaluation of the degree of horizontal anisotropy of a confined aquifer through following relation:

$$s = \frac{Q}{4\pi (KD)_e} W(u_{xy}),$$

where

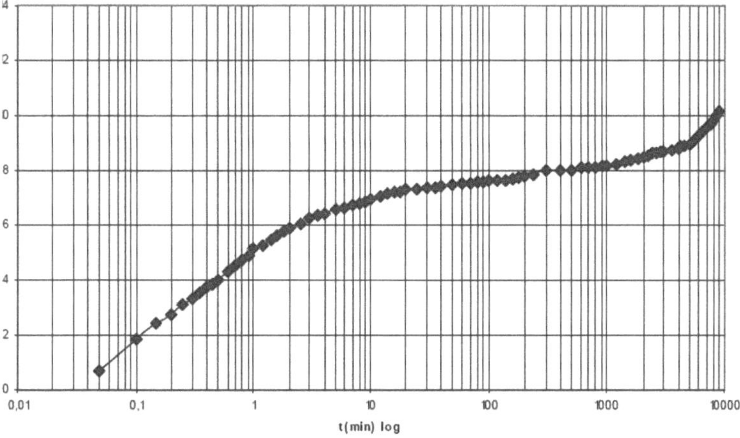

Fig. 2.15 Experimental curve that allows, together with numeric expressions, to obtain the main hydrogeological parameters

- $w(u)$ = "well function";
- $u_{xy} = \frac{r^2}{4t(KDt)_n}$ (Hantush, 1966);
- $(KD)_e = \sqrt{(KD)_x \cdot (KD)_y}$ = effective transmissivity;
- $(kD)_X$ = transmissivity in the direction with higher anisotropy;
- $(kD)_Y$ = transmissivity in the direction with lower anisotropy;
- $(kD)_n$ = transmissivity that considers the angles between the direction of the different piezometers of control with the axis X (higher anisotropy) (Fig. 2.16):

$$(KD)_n = \frac{(KD)_x}{\cos^2(\vartheta + \alpha_n) + m \cdot sen^2(\vartheta + \alpha_n)} \quad \text{and} \quad m = \frac{(KD)_x}{(KD)_y}.$$

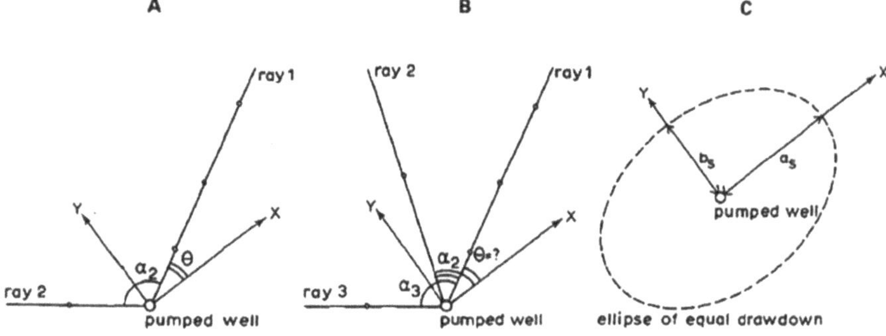

Fig. 2.16 Parameters used by Hantush for anisotropous aquifer on the horizontal plane (from Kruseman and de Ridder, 1994)

Using similar relations, following parameters can be determined:

- main direction of *horizontal anisotropy*, if it is not known beforehand, in a *confined aquifer* (Papadopulos, 1965; Neuman et al., 1984);
- the *vertical anisotropy* direction in a *confined aquifer* (Weeks, 1969);
- the *vertical anisotropy* direction in an *unconfined aquifer* (Neuman et al., 1984).

For in-depth studies, see the Bibliography.

Chapter 3
Influence of Joint Features on Rock Mass Hydraulic Conductivity

3.1 Introduction

As anticipated in previous chapters, rock mass hydraulic conductivity is highly influenced by physical and geometrical discontinuity features. In particular, the water flow inside a single discontinuity is ruled by aperture and roughness, whereas the rock mass hydraulic conductivity also depends on the frequency and persistence of joints. Those features are extremely variable in space due to the heterogeneity of the rock mass system. The definition of those relations apt to describe the variability of the parameters involved, both in a deterministic way and through a statistical approach, is therefore very important.

3.2 Influence of Joint Roughness

As discussed in Chapter 2, asperity height (ε) is among the necessary parameters for hydraulic conductivity calculation, in the case of rough joints.

The fact that fracture surfaces are not smooth, but show several and different asperities, has important consequences on the water flow inside fractures. The most evident effect is that the flow lines lose their parallelism (Louis, 1967); moreover, in comparison with a smooth joint with equal aperture, a rough joint involves a reduction of the hydraulic aperture and a lengthening of the hydraulic path, with a consequent increase in load losses. It is possible that all these factors can actually reduce hydraulic conductivity in the medium.

In the applied-geology field, in order to determine the hydraulic conductivity of a rock mass, the joint roughness becomes crucial only if the height of asperities is relevant when compared to the joint aperture. That means working generally in relative roughness $\varepsilon/D_h > 0.033$, corresponding to inertial flow.

L. Scesi, P. Gattinoni, *Water Circulation in Rocks*, DOI 10.1007/978-90-481-2417-6_3,

3.2.1 Effects of Roughness on Hydraulic Conductivity of a Single Joint: Theoretical Analysis

Bandis et al. (1985) proposes thefollowing empirical relation (Fig. 3.1), expressed in an implicit form, which allows the correlation of the JRC value (see Sect. Section 1.3.5) with the effective aperture of the joints, equivalent to the aperture of a smooth fracture:

$$e_e = (\text{JRC})^{2.5}(e/e_e)^{-2}$$

where

- e_e= effective discontinuity aperture (μm);
- e = mechanical aperture of fractures (μm), intended as the geometrical distance between the surfaces of the fracture (Fig. 3.2);
- JRC = joint roughness coefficient.

If it is assumed that

$$e_e = e - \epsilon,$$

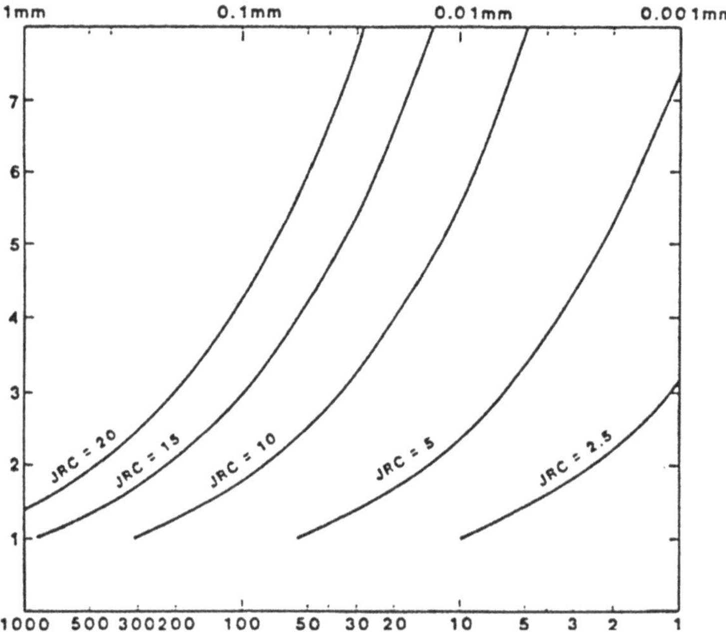

Fig. 3.1 Graphical representation of the results obtained using Bandis relation. The X-coordinate represents the real aperture (mm) and the Y-coordinate the ratio between real aperture and effective aperture. The different curves represent the different values of JRC considered (Lee and Farmer, 1993)

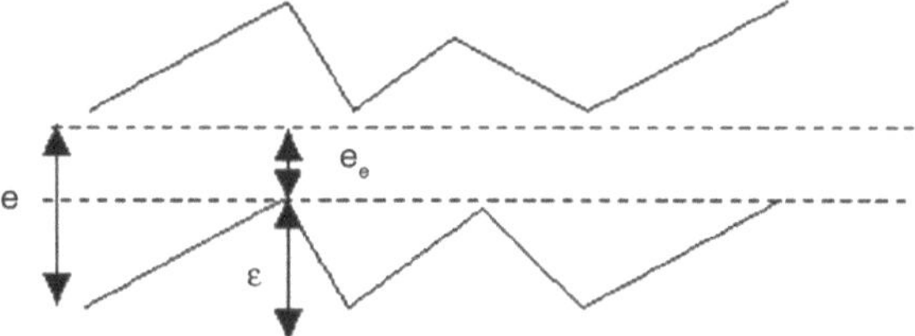

Fig. 3.2 Meaning of following magnitudes: "e" and "e_e" are respectively the mechanical aperture and the effective aperture of the discontinuity, whereas "ε" represents the asperity height

then

$$\epsilon = e - e_e = e - e^2/\mathrm{JRC}^{2.5} = e(1 - e/\mathrm{JRC}^{2.5}),$$

therefore, knowing that

$$D_h \cong 2e,$$

the result is

$$\frac{\varepsilon}{D_h} = \frac{e\left(1 - \frac{e}{JRC^{2.5}}\right)}{2e} = 0.5 - \frac{e}{2JRC^{2.5}}.$$

Combining that equation with the relations in Table 2.1, one can obtain the value K (referring to rough joints) according to the joint roughness coefficient JRC:

$$K = \frac{ge^2}{12v\left[1 + 8.8\left(0.5 - \frac{e}{2JRC^{2.5}}\right)^{1.5}\right]}, \tag{3.1}$$

($\varepsilon/D_h > 0.033$, inertial flow)

$$K = 4\sqrt{ge}\log\left(\frac{1.9}{0.5 - \frac{e}{2JRC^{2.5}}}\right), \tag{3.2}$$

($\varepsilon/D_h > 0.033$, turbulent flow)

$$K = 4\sqrt{ge}\log\left(\frac{3.7}{0.5 - \frac{e}{2JRC^{2.5}}}\right), \tag{3.3}$$

($\varepsilon/D_h < 0.033$, turbulent flow).

It is important to observe that the above stated equations include the Bandis empirical relation. For this reason the aperture (e) in the ratio $\frac{e}{2JRC^{2.5}}$ has to be expressed in μm, even if the ratio in itself has to be considered adimensional.

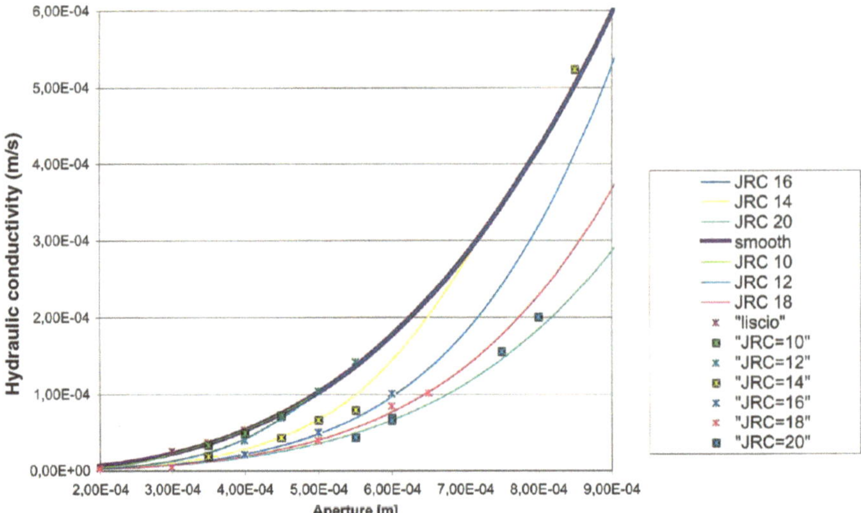

Fig. 3.3 Trend of hydraulic conductivity according to the aperture and the roughness coefficient using relation (3.1) valid for the non-parallel laminar condition

As it was predictable, the introduction of the roughness coefficient provides smaller conductivity values, with equal real aperture, according to the increasing roughness degree (Fig. 3.3); when the joint aperture increases, roughness has a more limited effect.

The validity of the above-mentioned expressions depends, as commonly known, on the Reynolds number and on the relative asperity height. Regarding particularly the laminar condition between rough joints, on the basis of the initial hypotheses ($\varepsilon/D_{h,}> 0.033$), thanks to which the relation (3.1) was obtained, it was possible to

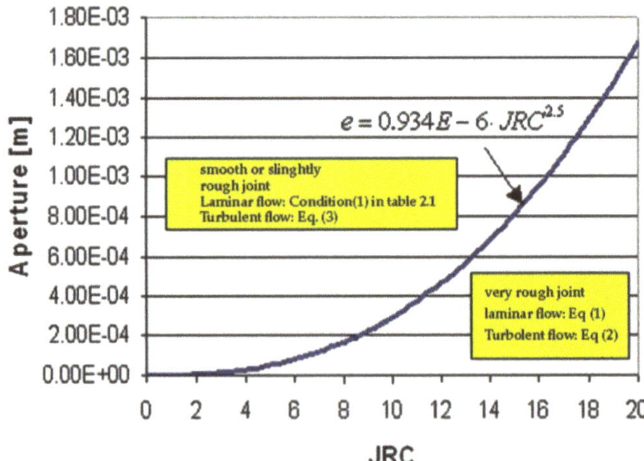

Fig. 3.4 Range of validity for relations (3.1), (3.2) and (3.3), for different flow conditions

identify, at a theoretical level, its existence field (Fig. 3.4) defined on the basis of the aperture and joint roughness coefficient (JRC) relationship:

$$e < 0.934 \times 10^{-6} \times \mathrm{JRC}^{2.5}$$

Therefore, it becomes even clearer that when the joint aperture increases, in laminar conditions, the roughness coefficient value must be raised in order to affect the fluid flow (Fig. 3.4). A max aperture value, $e_{\mathrm{JRC},}$ that can define the existence field of the different flow regimes, is therefore set.

For higher aperture values ($e \geq 0.934 \text{ x } 10^{-6} \text{ x } \mathrm{JRC}^{2.5}$) the fluid flow in laminar condition is not influenced by the joint roughness; therefore, it is possible to refer to the Snow expression for the hydraulic conductivity estimation. To sum it up, for profiles with not very marked asperities (JRC = 2, 4, 6), to allow a joint to be (from a hydraulic point of view) rough also in a laminar condition, a fracture's maximum aperture is almost 10^{-4} m; for profiles with intermediate roughness (JRC = 8, 10, 12, 14) the maximum aperture is between 10^{-4} and 10^{-3}m; finally, for profiles with high roughness (JRC = 16, 18, 20) the maximum aperture is about 10^{-3} m.

The above-presented theory assumes that the two joint faces which constitute the fracture present the same roughness profile; such a simplifying hypothesis is obviously not verified in real-life, as the two surfaces may have undergone a process of physical and chemical alteration which could have modified the shape or relative shifting of the asperities. In the first case, the problem can be easily solved by taking into consideration a mean JRC value between the ones associated with each joint face. In the second case, the effect of the asperities on the effective aperture of joints becomes greater. The most limiting hypothesis for down-flow estimation is that the joints must be in phase opposition; in such a case, it is necessary to refer to the following relation for the determination of relative roughness:

$$\frac{2e}{D_h} = 1 - \frac{e}{JRC^{2.5}}.$$

3.2.2 *Effects of Roughness on Hydraulic Conductivity of a Single Joint: Experimental Checking*

To experimentally verify the effect of roughness on a single joint (Gatttinoni and Scesi, 2004 and 2007), the necessary laboratory tests were undertaken. For this purpose, a suitable device for testing was specifically built which, based on the same working principle of the permeameter (Thiel, 1989), allows the evaluation of hydraulic conductivity along a joint of rock (Fig. 3.5).

The tests were carried out on cement samples modeled using Barton's profiles reproduced on wooden templates (Fig. 3.6). At a geologically applicative level, only the joints with high relative roughness have a remarkable importance; while, from a practical point of view, a joint having low roughness can be compared to a smooth joint. That approach allowed to verify mainly the validity of expressions (3.1) and (3.2), relative to inertial and turbulent flow (very rough joints).

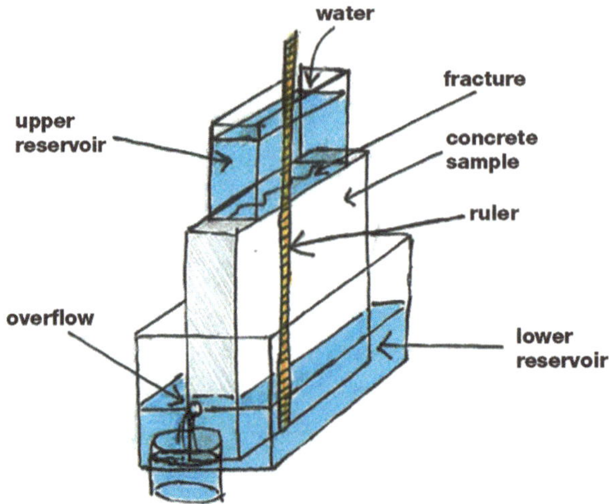

Fig. 3.5 Scheme of the device used for laboratory tests to measure the hydraulic conductivity

The hydraulic conductivity tests were executed under variable load (Fig. 3.7), evaluating conductivity as (Lambe and Whitman, 1969; ASTM D5084)

$$K_{sp} = \frac{A_T \cdot L}{A_C \cdot \Delta t} \ln\left(\frac{h_1}{h_2}\right),$$

where

- A_T = surface of the fractured sample;
- A_C = surface of the lower reservoir (downstream);
- h_1= water height in the upper reservoir (upstream) at time t_1;
- h_2= water height in the upper reservoir (upstream) at time t_2;
- L = length of the sample interested by the flow;
- $\Delta t = t_2 - t_1$ = time interval (30 seconds).

The Reynolds number (R_e) obtained by the different tests carried out on samples with quite elevated relative asperity height ($0.04 < \varepsilon/D_h < 0.4$) has always been smaller than 200, to indicate inertial non-totally turbulent flow conditions (for very rough joints, Louis indicate the critical value R_e in between 100 and 1000).

According to the results obtained (Fig. 3.8), the laboratory tests highlighted a fairly good adaptation of experimental data to theoretic curves, even though that check involved only the laminar flow and roughness coefficient JRC $\geq$ 12. Despite such limitations, the obtained result has a certain applicative interest, given that in nature, fractures which develop preferential flow paths generally have small apertures and roughness of a comparable order to that of the aperture itself; therefore, the observable flows are almost always inertial.

Only in presence of big karst bonds or fractured zones with particularly large apertures, the flow assumes a turbulent condition; in such cases, however, the joints'

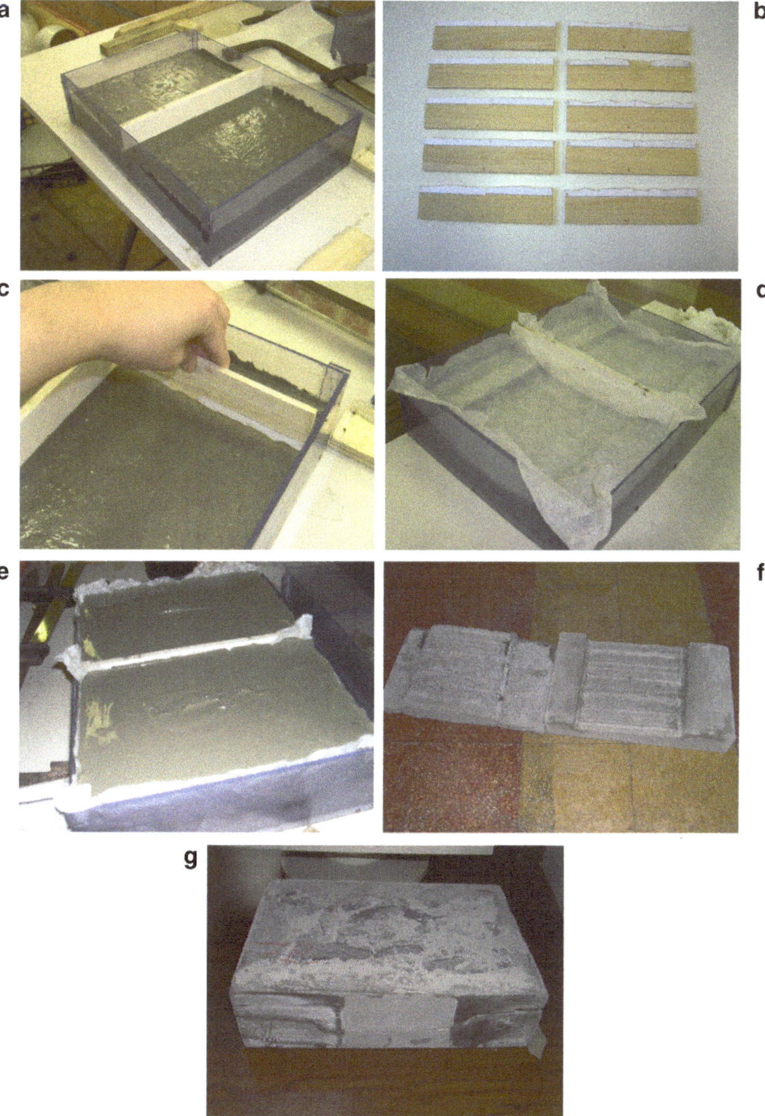

Fig. 3.6 Preparation procedure for tests samples to be used for the laboratory tests: (**a**) First cement layer casting; (**b**) wooden templates reproducing the Barton's 10 roughness profiles; (**c**) roughness is obtained using the wooden templates; (**d**) setting of sheets to separate the cement layers ; (**e**) second cement layer casting; (**f**) both faces of the sample with rough surfaces; and (**g**) sealed sample ready to be tested

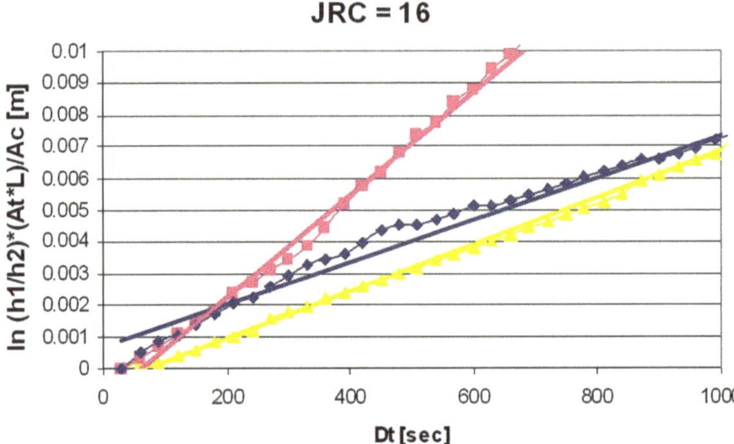

Fig. 3.7 Representative curves for hydraulic conductivity test under variable load; the example is referred to samples with JRC = 16. Conductivity represents the value of the tangent to the curves obtained: Steeper curves correspond to wider joint apertures and, therefore, to a higher conductivity

asperities are much smaller than the aperture. Therefore, the fluid flow is physically comparable to that typical of a rough pipe.

3.2.3 Effects of Roughness on Rock Mass Hydraulic Conductivity

Methodologies normally used to reconstruct the water flow in rock masses, starting from original formulations by Snow (1969), Louis (1974), Kiraly (1969), etc. up to the most recent ones (percolation theory, numerical modeling, etc.), generally provide overestimated conductivity values; the higher the asperities on the joint surfaces, the bigger the mistake. The problem in the application field becomes even more evident if it is considered that the rock mass is an anisotropous medium due to the orientation of discontinuities, which favour the flow in particular directions in space and prevent it in others; the presence of high roughness values along a discontinuity family may cause the "hydraulic closing", so that the flow along that direction is very scarce, thus influencing the trend of hydraulic circulation, often in a relevant way.

Let us take, for example, stability problems of slopes connected to water flow and the distribution of water pressures; in specific areas, the conductivity reduction due to the high roughness of the joints may actually prevent the water flow with the origination of water pressures, which may even be quite high.

When water cannot flow, it accumulates in fractures, and some blocks that are subject to the action of a destabilizing force of water can be interested by falls, topplings and/or slidings (Fig. 3.9).

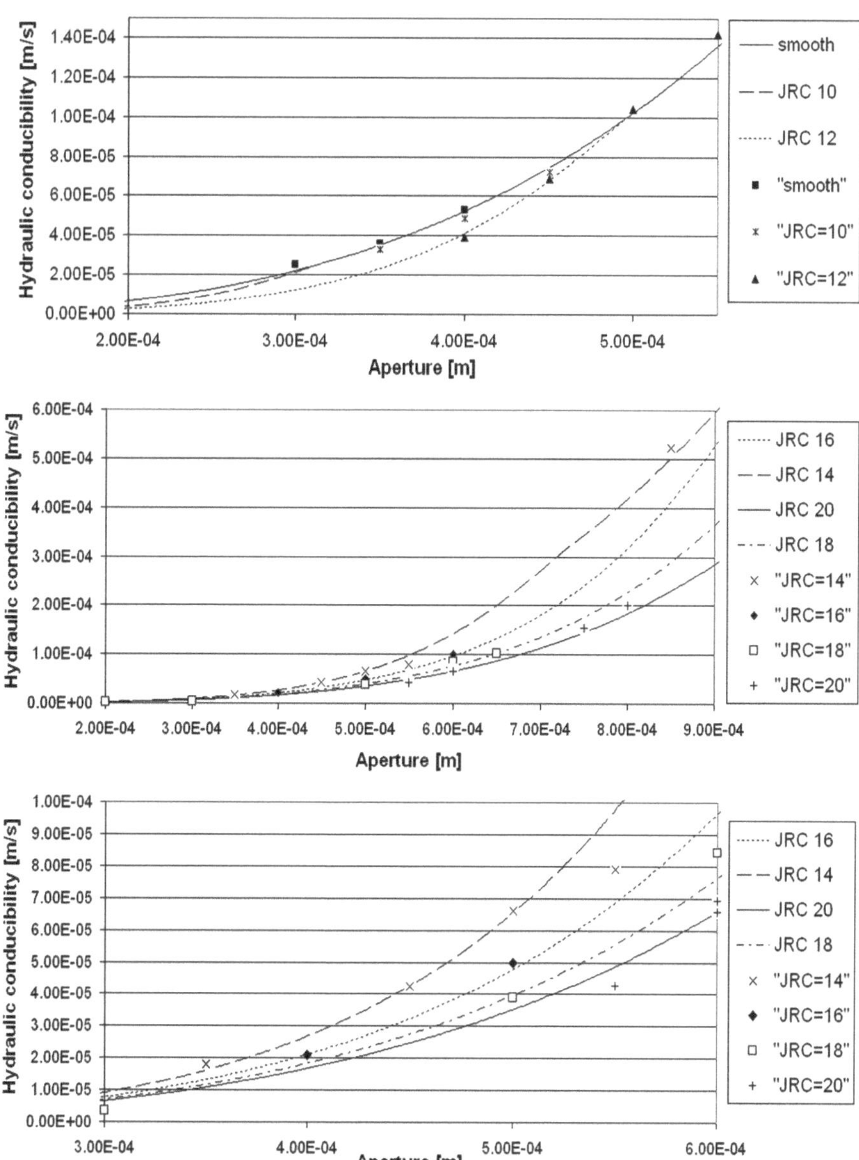

Fig. 3.8 Comparison among experimental (expressed with symbols) and the trend obtained from the equation for the laminar condition for different roughness coefficient JRC

Equally important are the consequences in the hydrogeological field, mainly in relation to problems connected to contaminants propagation in fractured aquifers. Actually, the introduction of the "roughness" parameters determines a further increase in the system complexity, with the formation of local piezometric anomalities that can influence the extension and the direction of contaminants that preferably

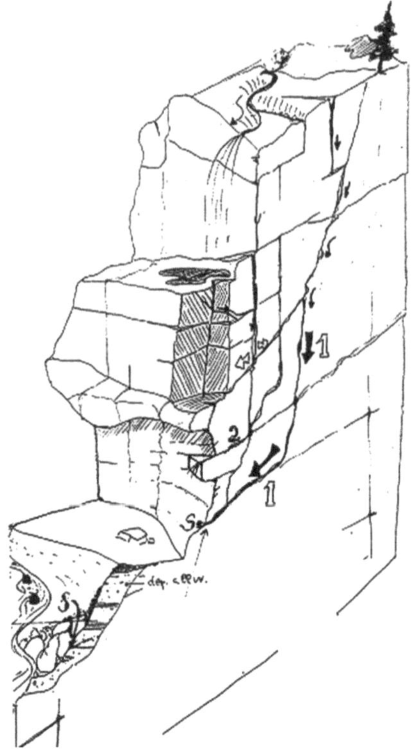

Fig. 3.9 Example of a slope with hydraulic conductivity (showed by the *arrows*), oriented preferentially along discontinuity "1", while discontinuities "2" (not so wide and characterized by higher roughness) tend to confine the water flow, thus originating water pore pressures and triggering potential instability phenomena (drawing by Francani)

propagate for advection, developing preferential paths, accumulations or dispersions according to the intersecting pattern of the different discontinuities. For example, let us consider a rock mass characterized by two discontinuity families: a vertical and a subhorizontal one; a higher roughness index for the subhorizontal discontinuity family determines a conductivity reduction in that direction, with a consequent reduction of the contaminant diffusion, associated with a contemporaneous increase in its trend to get deeper along the vertical discontinuity family.

3.3 Influence of Joint Aperture

The determination of conductivity in fractured rock masses can leave the knowledge of discontinuity aperture aside (see Chapter 2), as water can flow only along open joints. Nevertheless, the evaluation of the parameter "aperture" is very difficult, as it is influenced by various factors such as those given below:

1. *roughness*: It decreases the "effective aperture" of joints and, as a consequence, it influences water circulation (see Section 3.2);

2. *filling*: It is often present inside discontinuities and it can have different origins and nature. It can be made of calcite, chlorite, clay, silt, sand, coarse or fine material of tectonic origin etc. Therefore, according to the lithology and of the thickness (total or partial filling), the filling can cause important variations in the "hydraulic aperture" of the joint, as well as roughness;
3. *weathering*: When the discontinuity surfaces are very weathered (WD4; see Table 1.4), it is possible that there is an increase in the mechanical joint aperture. This does not always correspond to an increase in the hydraulic aperture, because it depends on the lithologic nature of the alteration material, that can favour or not the flowing of water;
4. *depth*: When the lithostatic load increases, discontinuities tend to close or reduce their initial aperture by far;
5. *karsts phenomena*: In those areas with relevant karsts phenomena, the discontinuity aperture can show local increases, also of different order of magnitude;
6. *anthropic action*: The use of explosives or of other excavation techniques used to build civil engineering facilities (tunnels, roads in mountain areas, etc.) can determine a reduction in the tension of the rock mass with a consequent increase in the joint aperture.

3.3.1 Changes in Aperture with Depth

There are empirical relations that allow to determine the mechanical and hydraulic aperture of discontinuities according to the depth starting from a mechanical initial aperture (Bandis et al., 1983, 1985) (Fig. 3.10):

$$e_t = e_{t_0} - \frac{0.027zV_m}{K_{ni}V_m + 0.027z} \quad \text{(mechanical aperture)},$$

where

- e_t = initial mechanical aperture;
- e_{to} = mechanical aperture at the considered depth;
- z = depth ($0.0027z = \sigma_v$ = lithosatic load);
- K_{ni} = initial stiffness: $0.02\left(\frac{JCS}{a_{t_0}}\right) + 1.75JRC - 7.15$;
- V_m = max discontinuity closure: $A + B(JRC) + C\left(\frac{JCS}{e_{t_0}}\right)^D$ with A, B, C, D = constants determined by cyclic load test; JRC = joint rough coefficient; JCS = compressive strength measured on the joint surfaces;

$$e = \frac{e_t^2}{JRC^{2.5}} = \frac{1}{JRC^{2.5}}\left(e_{t_0} - \frac{0.027zV_m}{K_{ni}V_m + 0.027z}\right)^2 \quad \text{(hydraulic aperture)}.$$

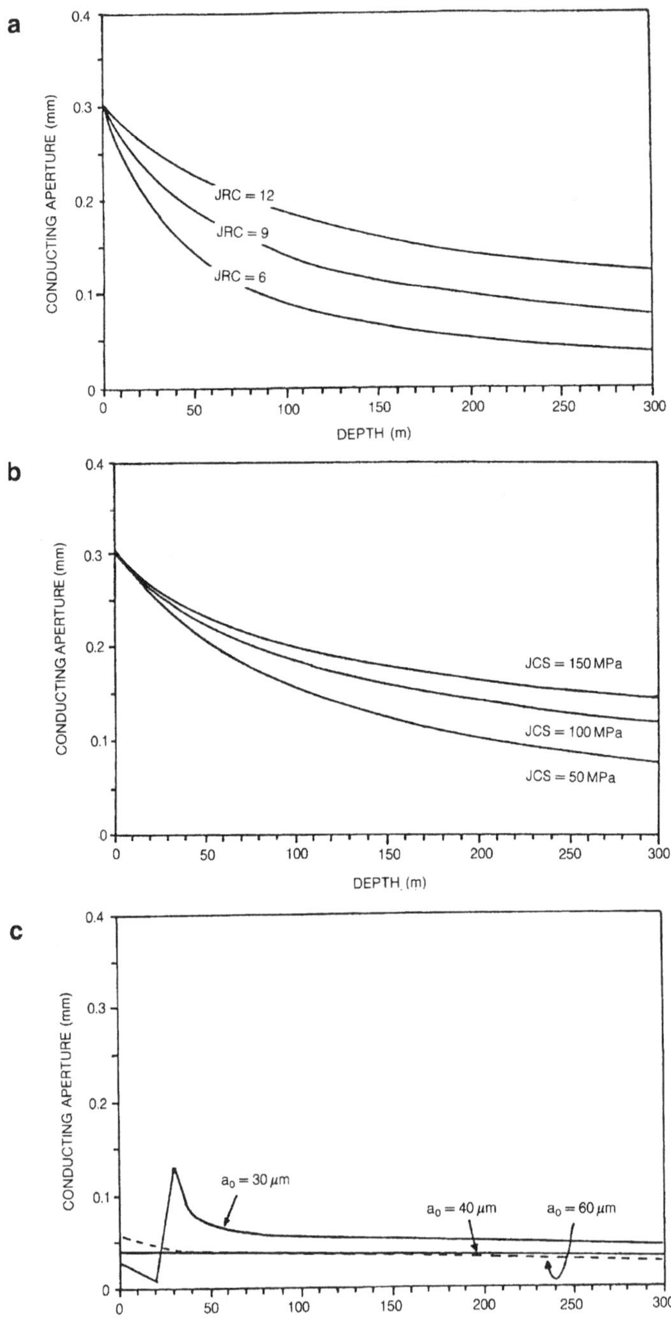

Fig. 3.10 Trend of joint hydraulic aperture with depth: (**a**) Effect of the changes of roughness coefficient JRC; (**b**) effect of the changes in the weathering degree expressed as joint compressive strength JCS; and (**c**) effects of the changes of superficial aperture e_0 (Lee and Farmer, 1993)

In order to know the variation of the joint aperture with depth, Lugeon tests can be made in a borehole at different levels.

In order to determine the *average aperture* of discontinuities in the area influenced by the test, the Cassan relation (1980) is generally used:

$$e = \left(\frac{6 \cdot Q \cdot \eta}{P_e}\right)^{1/3},$$

where

- e = average aperture (m);
- Q = absorption (m/s);
- η = water dynamic viscosity (10^{-4} kg s/m^2);
- P_e = effective pressure (kg/m^2).

Knowing the average aperture values for the different depths it is therefore possible to find the function that better approximates that change (Figs. 3.11, 3.12 and 3.13).

Based on what is stated, discontinuities would tend to close with depth, therefore the conductivity value should also decrease progressively. Some authors even

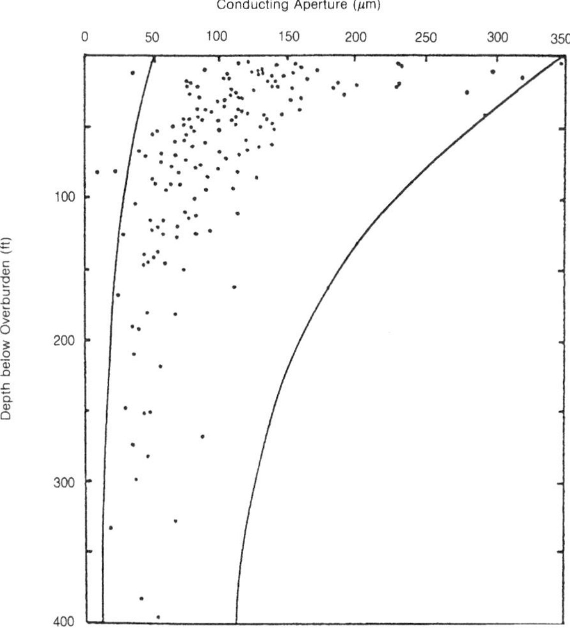

Fig. 3.11 Comparison among experimental data of hydraulic aperture in depth and theoretic curves (Lee and Farmer, 1993)

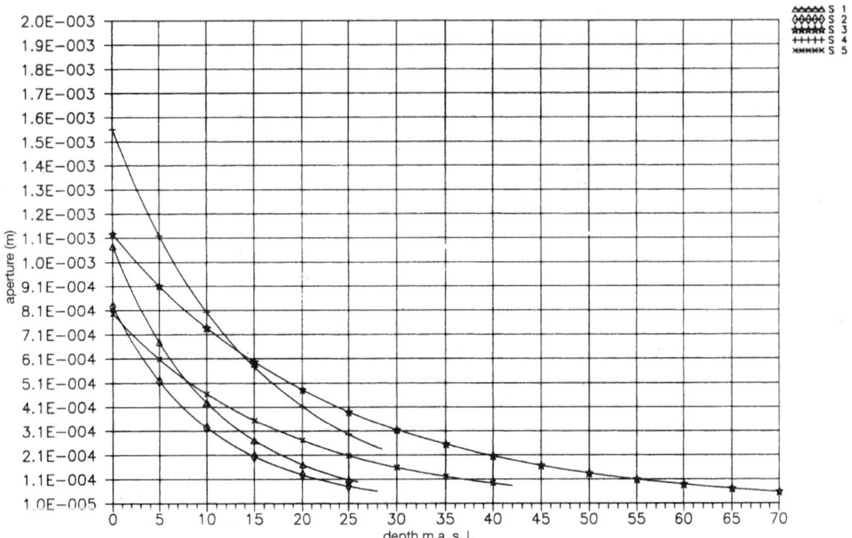

Fig. 3.12 Example of exponential variation of aperture with depth in five drilling tests near Lake Maggiore, on the Piedmontese side (Scesi and Saibene, 1989)

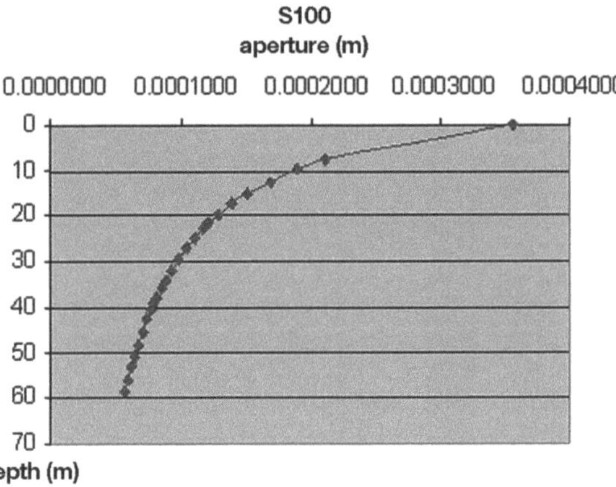

Fig. 3.13 Example of the trend of aperture with depth (flysch in Bergamo) obtained by the calibration made on conductivity tests at different depths

proposed empirical relations that allow to determine the conductivity according to depth Z:

$$\log \mathrm{K} = 5.57 + 0.352 \log \mathrm{Z} - 0.978(\log \mathrm{Z})^2 + 0.167(\log \mathrm{Z})^3 \quad \text{(Burgess,1977)}$$

where

- K = hydraulic conductivity (m/s);
- $Z = \delta_v/\gamma$ = depth (m).

That does not always correspond to reality, because the change in conductivity with depth does not only depend on the lithostatic load, but on other factors as well, such as lithologic changes, karsts phenomena, strengths, tectonic setting. Therefore, each time the importance to be attributed to each one of these factors must be evaluated according to the area examined.

3.3.2 Changes in Aperture with the Stress Field

A specific stress regime determines peculiar deformation structures in rocks, which are cinematically coherent and recognizable in the different observation scales (macro-, meso- and microscale) (Fig. 3.14).

Identifying the rock deformation is quite important for the effects they trigger; actually, rocks that underwent relevant deformation phenomena show poor technical features and are more permeable.

If a given area is only subject to extension or shortening displacement, but having different magnitude, the following effect can be observed: Open fractures (if siliceous or carbonatic solutions fill those fractures, "veins" will form) or stylolites of tectonic origin, with variable dip according to the entity and strike of the stresses (Fig. 3.15).

Instead, systems of peculiar fractures similar to those shown in Fig. 3.16 develop in the areas that are only subject to shear phenomena (stripe slip phenomena).

From the examination of discontinuities and structures that originate after tectonic events (veins, stylolites, striae etc.) it is therefore possible to go back to the strike of the stresses and to set the families characterized by a wider aperture and, as a consequence, by a higher conductivity.

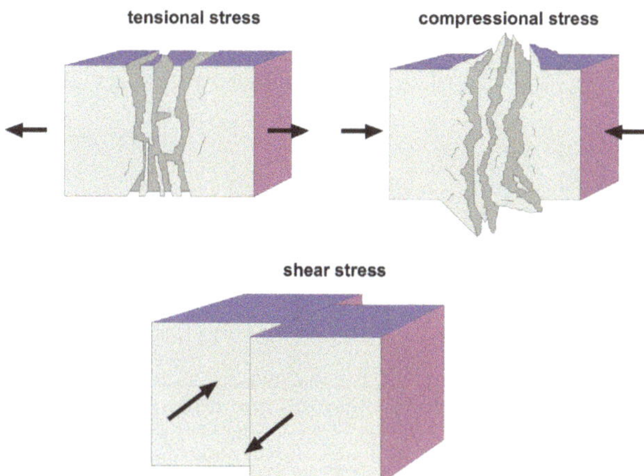

Fig. 3.14 Examples of stresses that can affect rocks

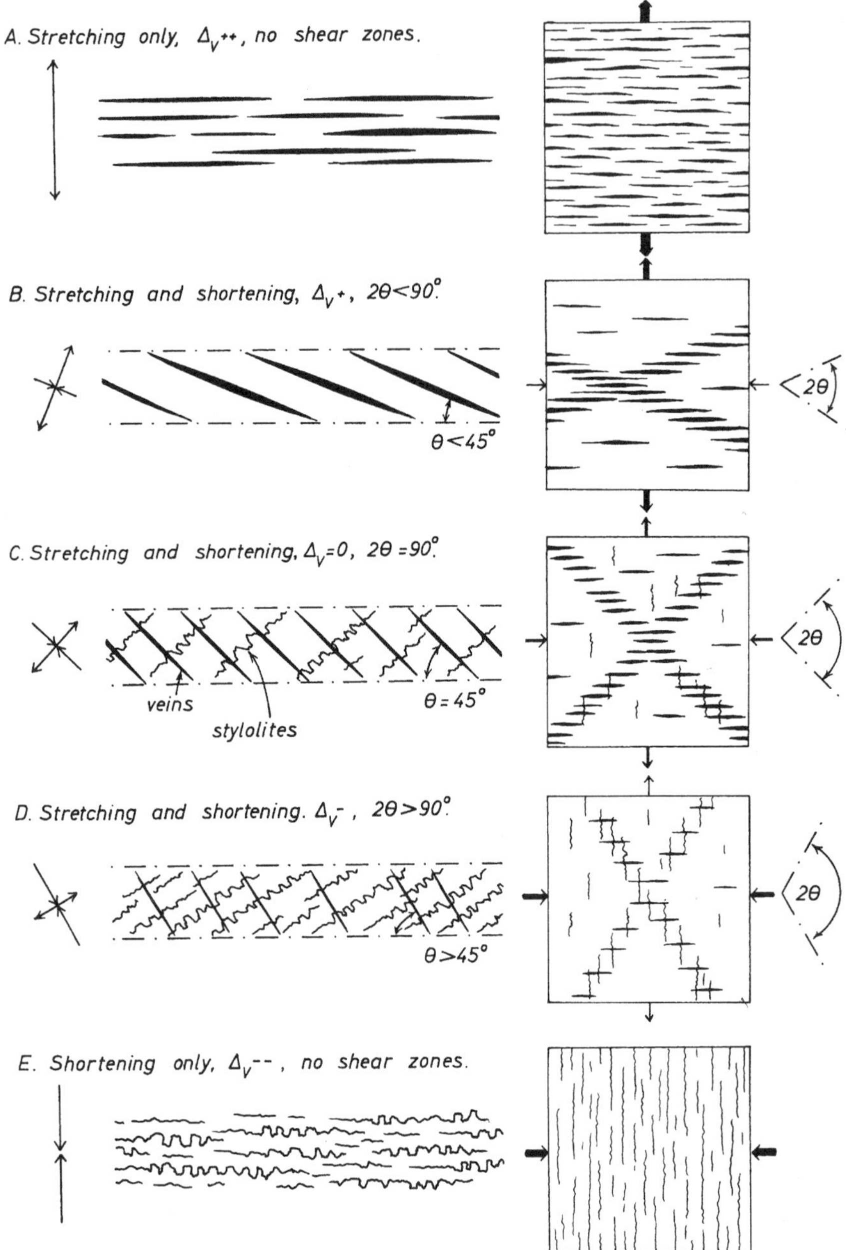

Fig. 3.15 Veins and stylolites that originate in rocks that were subject to compressive or tensile stress (Ramsay and Huber, 1987)

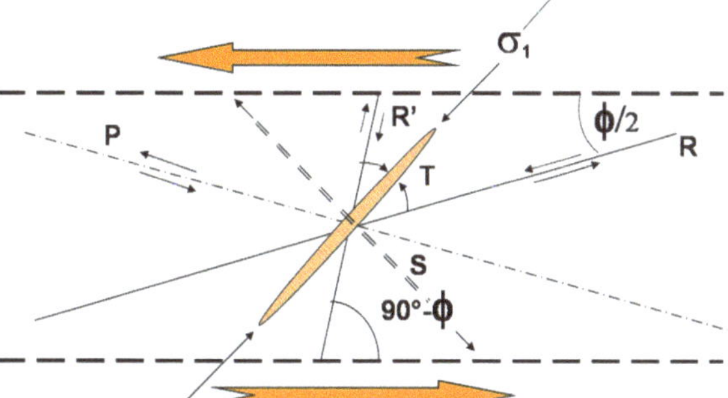

Fig. 3.16 Examples of fractures that originate in a shear zone (R and R′ = Riedel fractures; P = Skempton fracture; T = tension cracks; S = foliation; F = friction angle)

Considering the strike of the main stresses

- σ_1 coincides with the strike of max stresses;
- σ_2 lay on the normal plane of the movement strike (it can be determined from the exam of the tectonic structures);
- $\sigma_3$lay on the normal plane to that determined by the previous two.

These main stresses can be visualized very well on stereographic representations similar to those in Fig. 3.17.

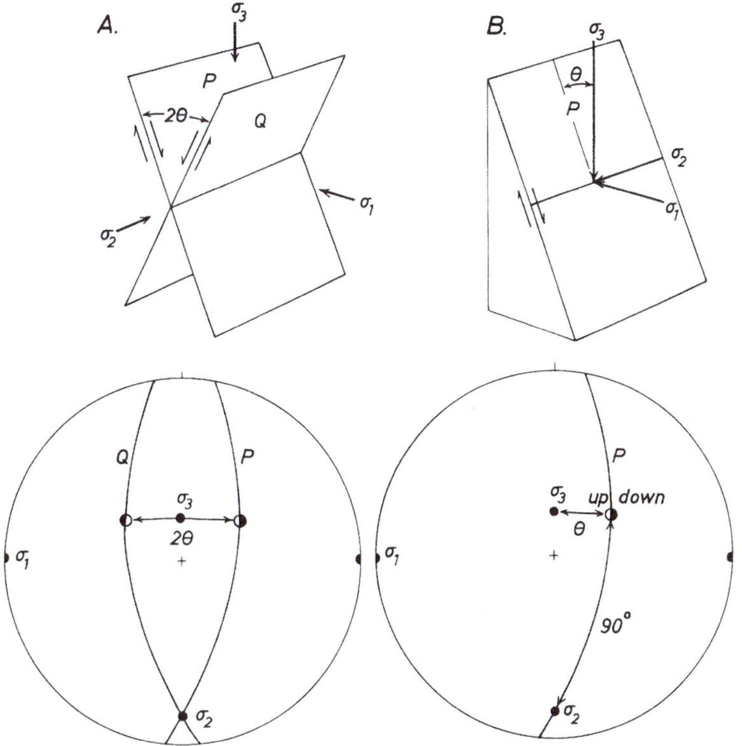

Fig. 3.17 Stereographic representation of the main stresses σ_1, σ_2, σ_3 (Ramsay and Huber, 1987)

It is evident that if the field of stress changes, the discontinuity aperture and the asperities present on the joint surface also change and, as a consequence, the conductivity of the medium and the flowing mode of the fluid vary.

Many authors tried to find out relations apt to determine the link existing between stress changes and, as a consequence, of the deformations undergone by rocks and changes in conductivity (Gangi, 1978; Walsh and Brace, 1984; Bai and Elsworth, 1994; Bai et al., 1999; Liu et al., 2000).

It is actually known that the boring of tunnels and mines or pits, the explosion of mines, the oil extraction, etc. imply changes in the stress field and, as a consequence, changes in the discontinuity aperture and/or spacing.

Usually, following expressions are used:

$$K = K_0 \left\{ 1 - \frac{1}{2} \left(\frac{\Delta\sigma}{E} \right)^{2/3} \right\}^4 \text{ (Gangi, 1978; relation true for soil),}$$

where

- K = final hydraulic conductivity;
- K_0 = initial hydraulic conductivity;
- E = elastic module;
- $\Delta\sigma$ = change in the stress field;

$$\frac{K}{K_0} = \left[1 + \frac{\Delta\varepsilon}{e} \left(\frac{k_n}{E} + \frac{1}{s} \right)^{-1} + \frac{\Delta\gamma}{e} \left(\frac{1}{s} + \frac{k_{sh}}{G} \right) (\tan \phi_d) \right]^3$$

(Bai et al., 1994 e 1999; relation valid for rocks),

where

- K = final hydraulic conductivity;
- K_0 = initial hydraulic conductivity;
- k_{sh} = tangential stiffness;
- k_n = normal stiffness;
- G = shear module;
- E = elastic module;
- s = joint spacing;
- e = joint aperture;
- ϕ_d = dilatancy angle;
- $\Delta\varepsilon$ e $\Delta\gamma$ = normal and tangential deformation;

$$\frac{K}{K_0} = \left[1 + \frac{2\,(1 - \mathrm{R_e})}{\varphi_f} \Delta\varepsilon + \frac{2\left(1 - R_q\right)}{\varphi_f} |\Delta\gamma| \right]^3$$

(Liu et al., 2000; relation valid for rocks),

where

- K = final hydraulic conductivity;
- K_0 = initial hydraulic conductivity;
- $\Delta\varepsilon$ and $\Delta\gamma$ = normal and tangential deformation;
- $\varphi_f \approx 2e/s$;
- $R_e = E/E_r$, con E = elastic module of the rock mass and E_r = elastic module of intact rock;
- $R_q = G/G_r$, con G = shear module of the rock mass and G_r = shear module of intact rock.

R_e and R_q can be correlated with Bieniawski's RMR quality index, that is

$$R_e = R_e = 0.000028\text{RMR}^2 + 0.009e^{\text{RMR}/22.82.}$$

3.4 Influence of Joint Spacing and Frequency

As already anticipated in Chapter 1, the fracturing degree of a rock mass is generally variable in space also according to the stresses undergone by the rock. In general, it can be stated that

- superficial portions of rock masses are interested by a higher number of discontinuities (reduced spacing and high frequency), as they are more exposed to the action of atmospheric agents and to weathering;
- the fracture degree is very high in correspondence to tectonized areas (faults, overthrusts) or fold structures (anticlinal or synclinal folds).

Statistical techniques are generally used to take into account the variability of the fracturing degree; they consist of making statistical tests on samples of homogeneous data and in defining, consequently, the more suitable probability distribution (Fig. 3.18).

As the hydraulic conductivity of a rock mass also depends on the fracturation intensity, it is variable in the different zones of the same rock mass, also due to the changes with depth that were experimentally observed by many authors (Snow, 1970; Louis, 1974; Gangi, 1978; Walsh, 1981).

If in boreholes used for the Lugeon tests, RQD are available as well, the *average frequency* of the discontinuities present at various levels can be obtained using Priest's and Hudson's relations (1976):

$$\text{ROD} = 100\,(0.1f + 1)e^{-0.1f};$$

$$\text{ROD} = 115 - 3.3J_v,$$

where

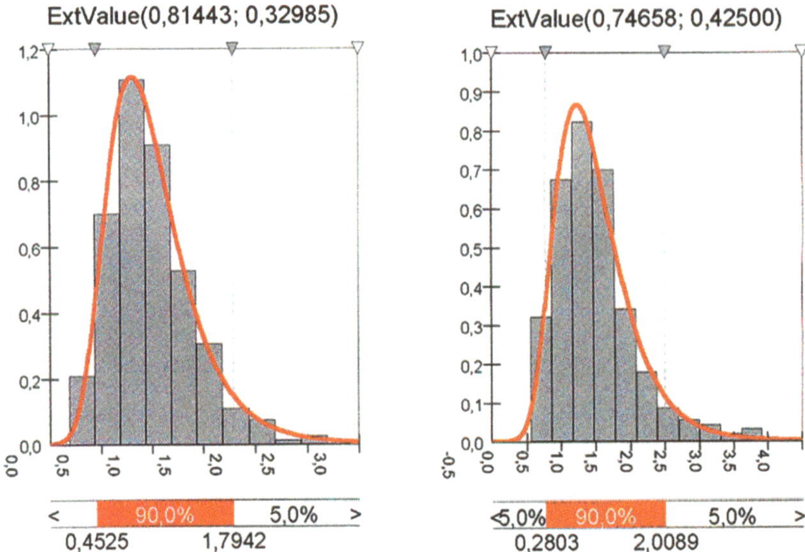

Fig. 3.18 Example of statistical distribution of spacing (**a**) and intercept (**b**) values in correspondence of a shear zone (Gattinoni et al., 2001)

- f = frequency (number of discontinuities per meter);
- J_v = number of discontinuities per volume unit.

By knowing the apertures and frequencies at the different levels, the change of conductivity with depth can be obtained (Fig. 3.19).

This kind of approach usually also allows to obtain a sufficiently large sample of conductivity data that can be used for the statistical tests; in order to guarantee the good quality of the data sample, it is appropriate to divide it according to the lithologic-structural setting, so as to obtain better results from the tests.

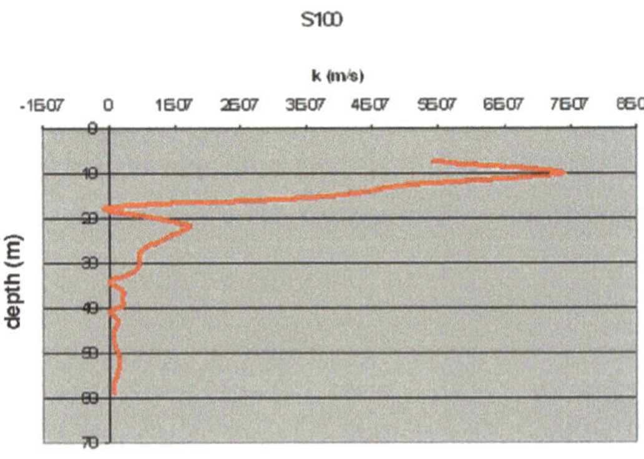

Fig. 3.19 Example of conductivity trend with depth (flysch in Bergamo)

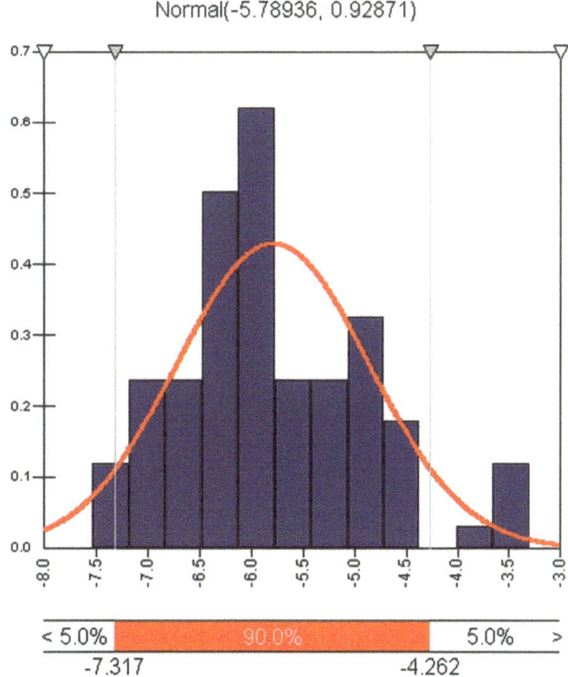

Fig. 3.20 Normal distribution of probability that is particularly suitable to describe the hydraulic conductivity logarithm in fractured rock, with a normal distribution

By way of example, the results of a sample of conductivity data referring to flysch are shown, for which the Generalized Beta proved to be the distribution that better approximates the sample (*p-value* 70% and 50% $< \alpha <$ 75%) (Fig. 3.20), and a sample of data referring to conglomerates in a fractured zone, for which the normal distribution resulted in the more suitable probability distribution.

3.5 Joints Interconnection

A further aspect to consider in reconstructing the preferential direction of the flow is the degree of interconnection among the different discontinuities (Fig. 3.21); actually, if fractures are independent from each other, they behave as isolated bonds that do not contribute to the shift of the fluid mass. If, on the contrary, the discontinuities are interconnected, conductivity is influenced by all discontinuity families present in the rock mass and the water flow follows a real flow network formed by a series of bonds connected to each other (Fig. 3.22). Keeping into consideration the interconnection when studying the percolation within a fractured rock mass means considering the probability that two or more discontinuity families may intersect.

Long and Witherspoon (1985) highlighted the fundamental importance of discontinuity interconnection for the conductivity of a rock mass both in terms of strike and module.

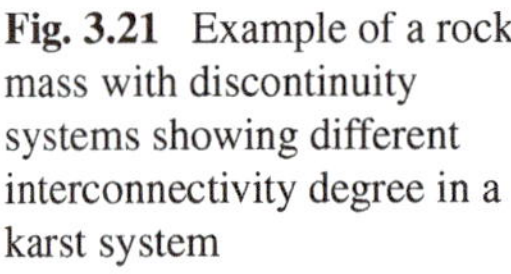

Fig. 3.21 Example of a rock mass with discontinuity systems showing different interconnectivity degree in a karst system

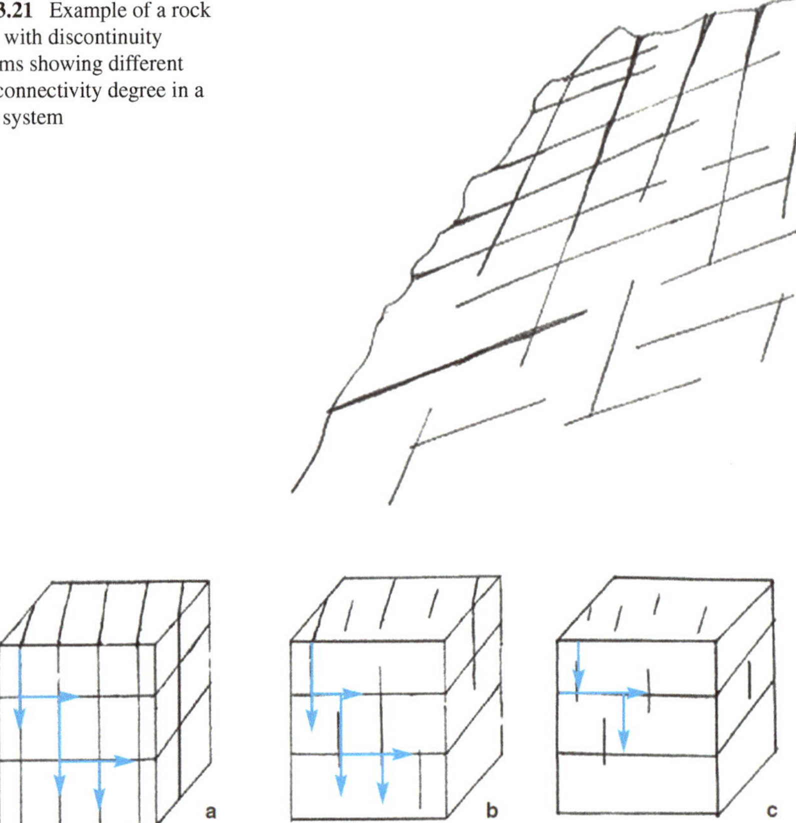

Fig. 3.22 Tridimensional representation of fracture systems with different interconnectivity degree: (**a**) completely interconnected; (**b**) partially interconnected; and (**c**) not interconnected

Rouleau and Gale (1985) proved that interconnection (I_{ij}) between two discontinuity families may be expressed as a function of other parameters characteristic of the same discontinuity families, such as orientation, spacing and persistence:

$$I_{ij} = \frac{l_i}{s_i} \sin \gamma_{iJ}, \quad i \neq j$$

where l_i and s_i arethe length and the average spacing of the ith discontinuity family, respectively, whereas γ_{ij} is the angle between the two discontinuity families. In general, it happens that

$$I_{ij} = I_{ij}.$$

If n represents the number of discontinuity families inside the rock mass, the comprehensive interconnection of the ith discontinuity family is given by

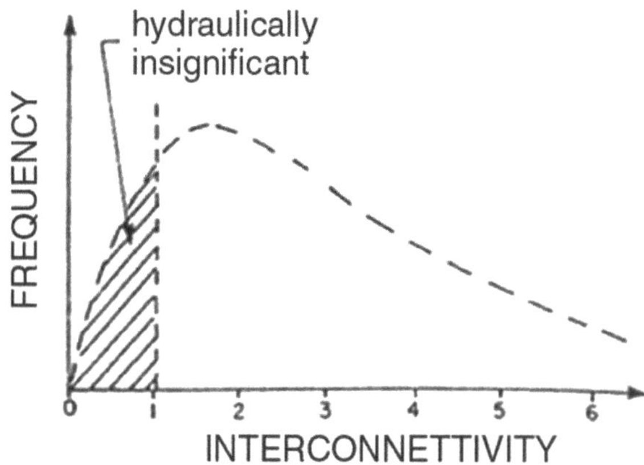

Fig. 3.23 Example of interconnectivity function relative to a discontinuity family (Lee and Farmer, 1993)

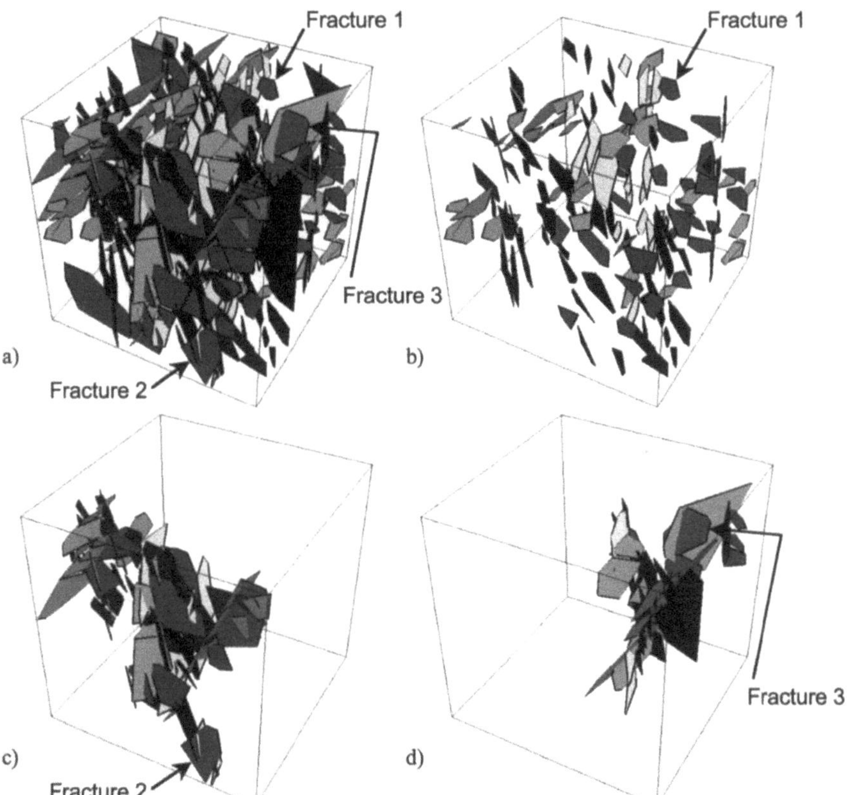

Fig. 3.24 Schematization of the clustering process of fractures in sub-networks according to the geologic-stochastic approach (Meyer and Einstein, 2002); (**a**) comprehensive fracture network; (**b**) isolated fractures; and (**c**) subsystems of larger interconnected fractures

$$I_i = \sum_{j=1}^{n} I_{ij}, \quad i \neq j$$

Naturally, when I_i increases, the load of the ith discontinuity family increases from the hydraulic point of view. The unit (Fig. 3.23) is assumed as the limit value of the interconnection index I_i:

- if $I_i > 1$, the ith discontinuity family is interconnected with the rest of the discontinuous system;
- if $I_i < 1$, the ith discontinuity family is isolated from the rest of the discontinuous system.

The percolation theory (see Chapter 2) is a probabilistic method that takes into account the interconnectivity among discontinuity; if the frequency and persistence values of the different discontinuity families are known, it allows the evaluation of an intersection probability among the fractures, the so-called connectivity (ξ). On the basis of the connectivity, it is possible to define the probability p that two discontinuities meet, and a critical interconnectivity probability p_c (see Sect. 2.3), so that the kind of percolation is defined

- if $p > p_c \Rightarrow$ interconnected flow;
- if $p < p_c \Rightarrow$ localized flow.

Recent models for the generation of fracture systems also simulate the interconnectivity with a stochastic approach (Meyer and Einstein, 2002) and they allow to forecast the water flow within the rock mass according to the connectivity degree (Fig. 3.24).

Chapter 4
Main Flow Direction in Rock Masses

4.1 Introduction

Aquifers, and aquifers in rock masses in particular, present a relevant heterogeneity in the study scale used to solve application problems (slope stability, road and tunnel construction, etc.). The presence of those heterogeneous features, often combined with a marked anisotropy of the medium, implies the need of a careful reconstruction of the hydrogeological structure, so to make the conceptual model acceptable and the forecast of the water flow reliable enough; usually, the waterflow is ruled by the orientation and the hydraulic features of the joints.

Actually, in rock masses, water flows along paths that do not always coincide with that of the piezometrical gradient; on the contrary, this usually happens in homogeneous and isotropic media. When the medium presents a high hydraulic conductivity in a certain direction, the flow is highly conditioned by this anisotropy and it tends to occur as to favour the mass transfer in that direction.

Similar effects disclose in the presence of impermeable layers that, preventing the flow along the direction orthogonal to their surface, determine the flow orientation that almost coincides with the above mentioned impermeable layer.

This chapter will analyze the waterflow conditions inside rock media, both in saturated and non-saturated (percolation flow) conditions, and it will present an easy to apply procedure to determine the main flow direction, also in relation to the presence of peculiar hydrogeological structures. In particular, starting from the structural discontinuity features (see Chapter 1), it is possible to determine the hydraulic conductivity tensor (see Chapter 2), its components in the different directions of the hydraulic gradient and the main percolation direction, also in presence of an impermeable bedrock. Moreover, this information can be plotted using stereographic representations.

4.2 Anisotropy of the Fractured Medium

The marked hydraulic conductivity orientation typical of rock masses (anisotropous media) strongly affects the groundwater flow; for example, the presence of impermeable layers preventing the flow in the direction orthogonal to their surface

L. Scesi, P. Gattinoni, *Water Circulation in Rocks*, DOI 10.1007/978-90-481-2417-6_4,

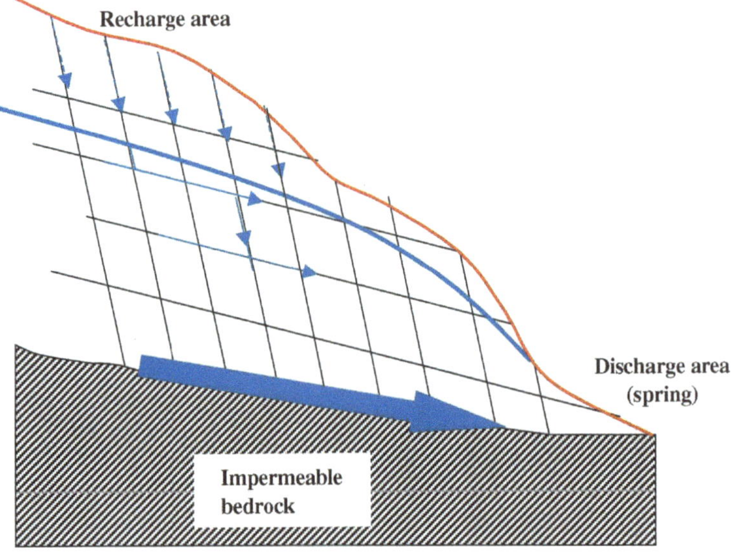

Fig. 4.1 Schematic representation of the flow condition typical of a rock slope with an impermeable bedrock, with indication of the water flow direction within the discontinuity network. The water path is conditioned by the hydraulic gradient and by the discontinuity distribution and orientation

determines a flow that almost coincides with the impermeable layer itself (Fig. 4.1). For a quantitative evaluation of the phenomenon, the setting of an *anisotropy vector* K_r is very useful:

$$K_r = |K_1/K_3 K_1/K_2 K_2/K_3|,$$

where K_1, K_2 and K_3 are the main hydraulic conductivity values of the tensor.

This anisotropy vector shows the relation among the hydraulic conductivities along the different directions in space.

That anisotropy vector is particularly significant in presence of shear zones (zone with high deformation with limited depth compared to their longitudinal extension; Ramsay and Huber, 1987), where, potentially, the main orientation of water flow occurs. If the hydraulic conductivities along the main directions (corresponding to the minimum, medium and maximum permeability) are known, it is possible to verify if they coincide with the gradient. In particular, if the rock shows no hydraulic conductivity in the gradient strike, the water flow is prevented, even though the overall permeability of the rock mass is different from zero.

From the examination of the hydraulic conductivity tensors and, as a consequence, of the anisotropy vector, it is possible to identify the main flow direction, to delimit the hydrogeological catchment and to identify the charge and discharge areas of the aquifer, as well as to provide the necessary data to quantify the flow.

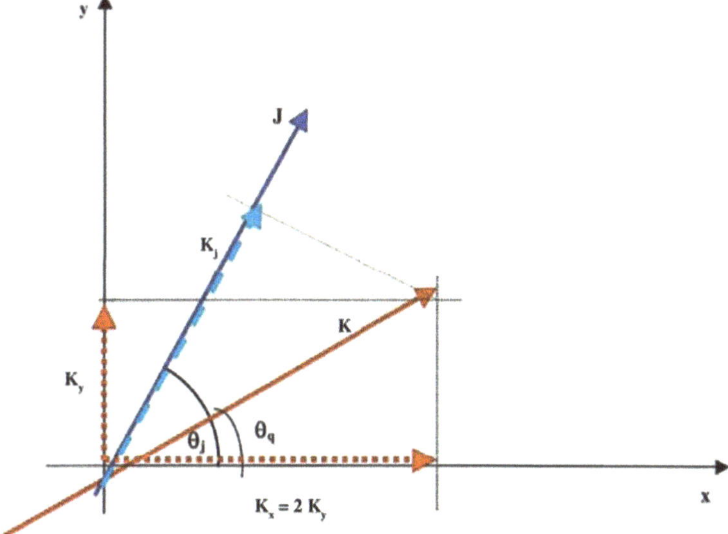

Fig. 4.2 The *red arrow* shows the main flow direction of the anisotropous medium, characterized by $K_x = 2K_y$. The hydraulic conductivity in the gradient direction K_J is shown in *blue*; in terms of module, $K_J < K$

In order to better understand what is stated above, it is useful to examine what happens in a bidimensional field. In that case, the relation that better fits the max hydraulic conductivity direction (θ_q), as a function of the anisotropy ratio in the plane considered (K_r), is (Fig. 4.2) given below:

$$\theta_q = \text{arctg}(K_r)$$

With a gradient different from θ_q, the hydraulic conductivity of the medium in the direction set by the gradient is lesser than the maximum one, thus reducing the flow as well.

In order to justify this expression, it is very important to understand along which direction the main mass fluid shift occurs in anisotropous media (*main flow direction*), and what follows has to be considered:

a. in saturated conditions, water tends to flow under the gradient impulse J and in not saturated conditions under the effect of gravity;
b. if the medium is cracked, water flows within bonds, following a "stepped" path, so that the resultant of the conditions along the different strikes leads the particles to respect, on average, the flow direction given by the gradient (in a condition of complete saturation) or the strike of the max hydraulic conductivity (in a non-saturated condition);
c. the waterflow along the hydraulic gradient is ruled by the projection of the hydraulic conductivity components in the strike of the gradient, which is surely lower than the max hydraulic conductivity relative to the main flow.

The identification of the main flow is very important to evaluate the groundwater flowing in the system, still this kind of analysis requires an accurate reconstruction of the hydrogeological structure in order to detect the presence of highly permeable layers (such as faults and shear zones) and the orientation of any impermeable layers.

4.3 Main Flow Direction in Fractured Media

To determine the main flow direction in a fractured medium it is important to take following elements into consideration: The permeability of single discontinuity families, the orientation and the interconnection of each set, the different flow conditions (saturated or non-saturated) and specific hydrogeological structures (e.g., impermeable bedrock).

The procedure leading to the identification of the main flow direction is described in the following paragraphs. This procedure implies the integrated use of easy calculation programmes and stereographic representations.

In the description of the method, many references will be made to an example where following discontinuity sets (Table 4.1) are considered:

- $F1 \rightarrow 90°/65°$ characterized by hydraulic conductivity K_{F1};
- $F2 \rightarrow 300°/30°$ characterized by hydraulic conductivity $K_{F2} = 0{,}5\ K_{F1}$;
- $F3 \rightarrow 240°/50°$ characterized by hydraulic conductivity $K_{F3} = K_{F2} = 0{,}5\ K_{F1}$.

Table 4.1 (**a**) Features of the single discontinuity sets and relative hydraulic conductivity and (**b**) hydraulic conductivity tensor of the rock mass characterized by the above-mentioned discontinuity sets

a			Average aperture of the family (m)	Average frequency of the family (1/m)	Hydraulic conductivity modulus (m/s)
			e	f	$K_{F1} = \frac{e^3 fg}{12\nu}$
$F1$	Dip direction (°)	90	0.00126	0.1	1.635×10^{-4}
	Dip (°)	65			
$F2$	Dip direction (°)	300	0.001	0.1	8.175×10^{-5}
	Dip (°)	30			
$F3$	Dip direction (°)	240	0.001	0.1	8.175×10^{-5}
	Dip (°)	50			

b	Tensor (m/s)	dip direction (°)	dip (°)	K_{eq} (m/s)
K_1	3.05×10^{-4}	189	1	
K_2	2.25×10^{-4}	99	1	2.04×10^{-4}
K_3	1.24×10^{-4}	66	88	

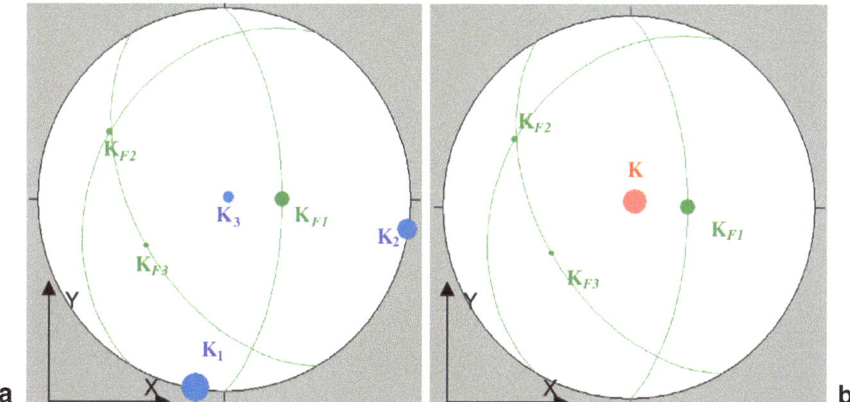

Fig. 4.3 (**a**) Stereographic representation of the hydraulic conductivity tensor in the rock mass (in *blue*) and of single hydraulic conductivity oriented along the discontinuity planes (in *green*); (**b**) stereographic representation of the main flow direction in a non-saturated medium: The size of coloured *circles*, representative of hydraulic conductivity vectors, is proportional to the modulus of hydraulic conductivity itself

The orientation of the discontinuity sets, the corresponding hydraulic conductivity values presented above and the rock mass hydraulic conductivity tensor (Table 4.1b and Fig. 4.3a) calculated through the relation in Chapter 2 can be plotted on the stereographic projection.

Whereas, to determine the main flow direction and the percolation rate, it is necessary to take the following data into account:

- the interconnection among the sets;
- the different flow conditions (saturated or non-saturated);
- the direction of the hydraulic gradient;
- the presence of impermeable layers.

4.4 Non-saturated Medium

In a non-saturated rock mass, the water flow inside discontinuities is ruled by gravity, therefore water tends to flow in depth following the orientation of discontinuities, especially the ones with high hydraulic conductivity (Fig. 4.3b). The water percolation goes on till it reaches either a watertable (in that case, a flowing condition occurs in the saturated medium that is ruled by the hydraulic gradient of the aquifer) or an impermeable layer that acts as a bedrock.

Therefore, the flow directions during the percolation phase coincide with the dip direction of each discontinuity family. The main flow direction is calculated by combining the discontinuity orientations, also considering the above mentioned flow components along the discontinuity sets (Fig. 4.4). To take into account the parameter *interconnection* those components can be considered according to the

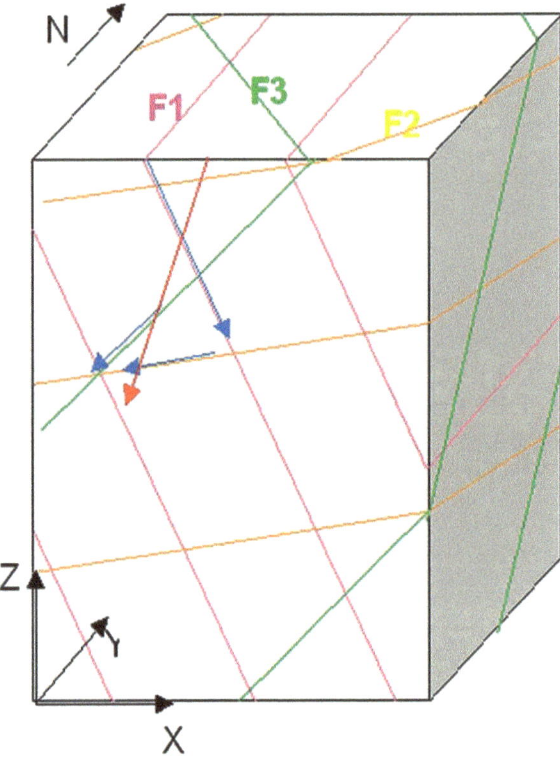

Fig. 4.4 Tridimensional view of the discontinuity sets associated to the three main percolation direction (in *blue*) and of the main flow direction (in *red*)

interconnection degree of each family with the system (Tables 4.2 and 4.3). The structural diagram of Fig. 4.3b shows the stereographic representation of each percolation direction and the main flow direction resulting from them.

Table 4.2 Interconnection degree among the different discontinuity sets: (**a**) Geometrical parameters used in the index evaluation and (**b**) interconnection matrix

a	Length (m)	Interconnectivity index normalized	Angle between the two planes (°)
	l	I	γ
	15	0.357	82.173
	15	0.331	48.597
	15	0.312	62.677

b	I_{ij}	$F1$	$F2$	$F3$	I_i
	$F1$	0	1.486	1.332	2.818
	$F2$	1.486	0	1.125	2.611
	$F3$	1.332	1.125	0	2.457

Table 4.3 Resultant of the composition of vectors relating to the percolation along the different discontinuity families: (**a**) Considering and (**b**) not considering the interconnectivity among the different discontinuity sets

a	Vectorial sum K_{tot}	
Vector comp.	x	-9.825×10^{-6}
	y	3.520×10^{-6}
	z	-8.597×10^{-5}
Modulus (m/s)	K_{tot}	8.660×10^{-5}
Dip direction	$\alpha\ (K_{tot})$	289.71°
Dip	$\psi(K_{tot})$	83.08°

b	Vectorial sum K_{tot}	
Vector comp.	x	-3.769×10^{-5}
	y	9.121×10^{-6}
	z	-2.516×10^{-4}
Modulus (m/s)	K_{tot}	2.546×10^{-4}
Dip direction	$\alpha\ (K_{tot})$	283.60°
Dip	$\psi(K_{tot})$	81.24°

Table 4.4 Features of the single discontinuity sets and relative permeability

			Average aperture of the family (m)	Average frequency of the family (1/m)	Hydraulic conductivity modulus (m/s)
			e	f	$K_{F1} = \frac{e^3 fg}{12v}$
$F1$	Dip direction (°)	270	0.001	0.1	8.172×10^{-5}
	Dip (°)	60			
$F2$	Dip direction (°)	180	0.001	0.1	8.172×10^{-5}
	Dip (°)	30			

Table 4.5 Main percolation direction: (**a**) Total interconnectivity; (**b**) partial interconnectivity of the discontinuity family $F2$ and (**c**) discontinuity family $F2$ completely isolated from the rest of the system

a	Length (m)	Normalized interconnectivity Index	Included angle (°)
	λ	I	γ
	20	0.500	64.341
	12	0.300	90

a	Vectorial sum K_{tot}	
Vector comp.	x	-4.086×10^{-5}
	y	-7.077×10^{-5}
	z	-1.116×10^{-4}
Modulus (m/s)	K_{tot}	1.383×10^{-4}
Dip direction	$\alpha\ (K_{tot})$	210.00°
Dip	$\psi(K_{tot})$	53.79°

b Length (m)	Normalized interconnectivity Index	Included angle (°)
λ	I	γ
20	0.625	64.341
12	0.375	90

b	Vectorial sum K_{tot}	
Vector comp.	x	-2.554×10^{-5}
	Y	-2.654×10^{-5}
	z	-5.955×10^{-5}
Modulus (m/s)	K_{tot}	7.002×10^{-5}
Dip direction	$\alpha\ (K_{tot})$	223.90°
Dip	$\psi(K_{tot})$	58.27°

c Length (m)	Normalized interconnectivity Index	Included angle (°)
λ	I	γ
20	1.000	64.341
10	0.000	90

c		
Vector comp.	x	-4.086×10^{-5}
	y	-7.509×10^{-21}
	z	-7.077×10^{-5}
Modulus (m/s)	K_{tot}	8.172×10^{-5}
Dip direction	$\alpha\ (K_{tot})$	270.00°
Dip	$\psi(K_{tot})$	60.00°

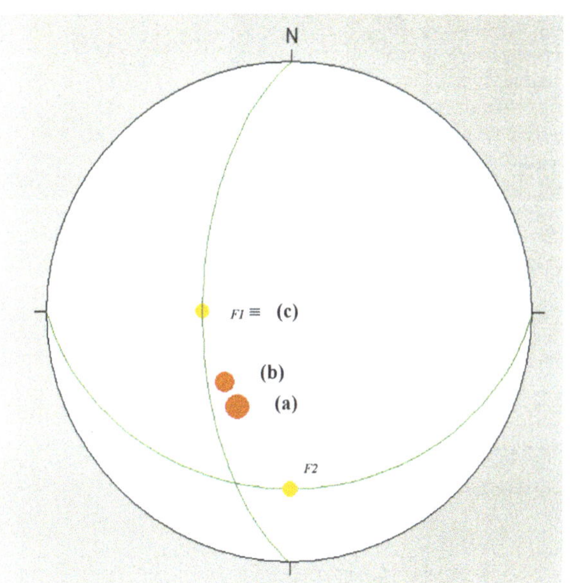

Fig. 4.5 Stereographic representation of the percolation directions in the two discontinuity sets (*F*1 having hydraulic conductivity K_1 and *F*2 having permeability K_2) and of the main percolation direction resulting from them for the different interconnectivity degrees: complete interconnection; (**b**) discontinuity *F*2 partially interconnected with the rest of the system and (**c**) isolated discontinuity *F*2 n (in *red*)

In the illustrative case presented in Table 4.3a, b, the role played by interconnectivity in determining the main percolation flow is evident; the influence of that parameter is even more evident when considering the extreme case of two discontinuity families (Table 4.4) characterized by a highly variable interconnection degree (Table 4.5). In that case, the percolation direction changes according to the interconnection index until it coincides with the orientation of the discontinuity families characterized by a higher persistence (Fig. 4.5).

4.5 Non-saturated Medium: Main Flow Direction with an Impermeable Layer

A peculiar situation occurs when the percolation water reaches an impermeable bedrock. Usually, that bedrock can be assimilated to a plane, therefore it is represented on the structural diagram with a great circle or a pole that shows hydraulic conductivity direction null, where the flow is prevented. The main flow direction can be determined graphically by also representing on the stereonet the great circles of the discontinuity families that define the main percolation directions. Actually, the flow direction is determined by the intersection between the great circle of the impermeable level and the great circles of the different discontinuity families (Fig. 4.6); the sum of the vectors thus obtained, whose module is equal to the hydraulic conductivity value along the corresponding discontinuity set, provides

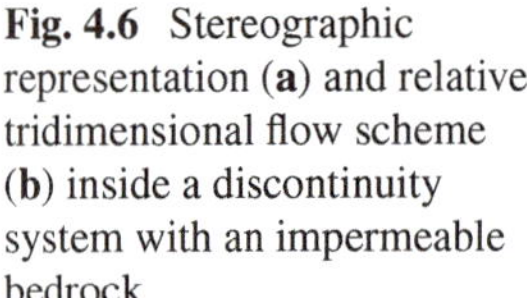

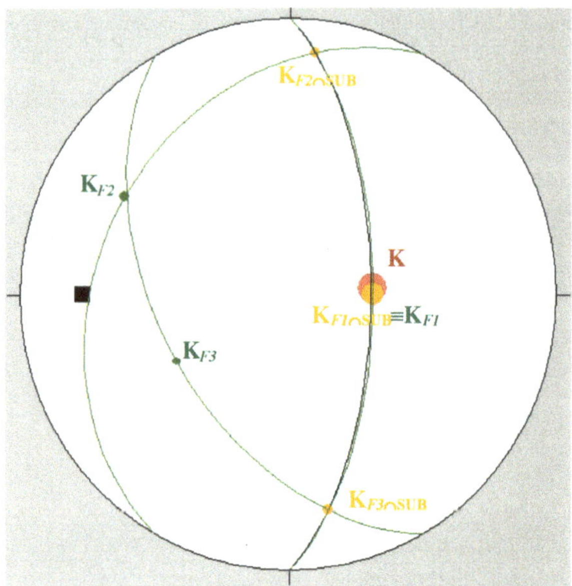

Fig. 4.6 Stereographic representation (**a**) and relative tridimensional flow scheme (**b**) inside a discontinuity system with an impermeable bedrock

the main flow direction (in red in the figure) determined by the presence of the impermeable layer (Table 4.6).

When, on the stereonet, the pole of the bedrock coincides with the percolation direction of a discontinuity family, it means that the set is orthogonal to the bedrock; in this case, it contributes to the percolation, but not to the water flow along the bedrock itself (Fig. 4.7). Quite often, a more fractured rock layer is present at the contact point; this layer could direct the waterflow parallelly to the bedrock. This situation can be simulated by introducing a new discontinuity family whose orientation coincides with that of the bedrock (as in the previous example).

4.6 Saturated Medium

When percolation water reaches a saturated level characterized by a certain hydraulic gradient (represented in blue in the stereonets in Fig. 4.8), the water flow follows the gradient, but it is greatly conditioned by the orientation of discontinuity planes. For example, if there is a single discontinuity set whose orientation is orthogonal to the hydraulic gradient, the flow is prevented because the hydraulic conductivity in that direction is null. When the rock mass is saturated, the hydraulic conductivity tensor can be calculated. Therefore, in order to assess the water flow it is necessary to calculate the hydraulic conductivity in the gradient direction.

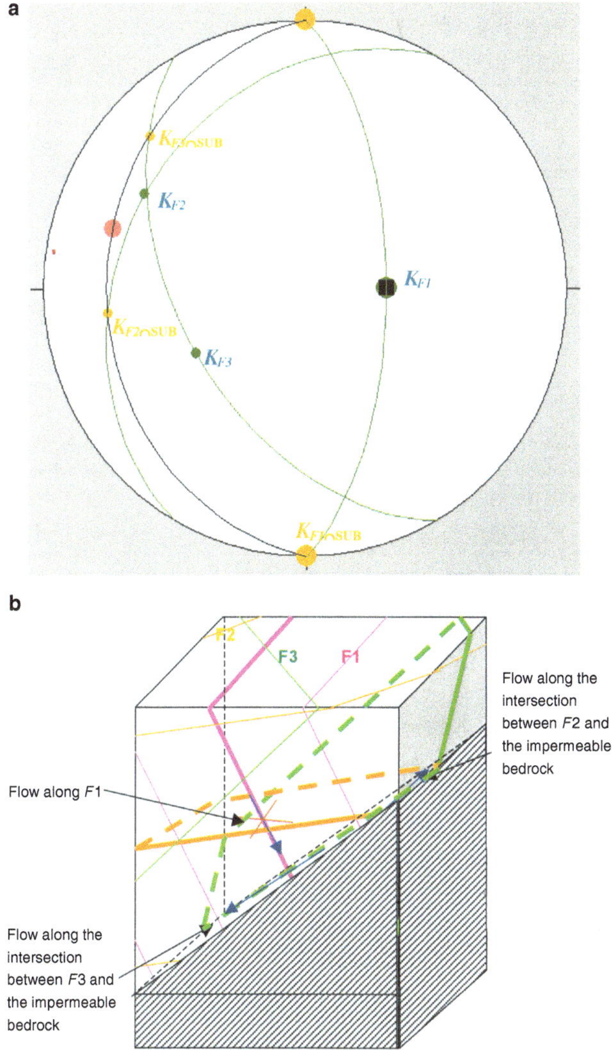

Fig. 4.7 Stereographic representation (**a**) and relative tridimensional flow scheme (**b**) inside a discontinuity system with an impermeable bedrock, orthogonal to one of the hydraulic conductivity vectors

4.6.1 Known Hydraulic Gradient

On the stereonet, the hydraulic gradient can be shown as a plane, characterized by dip direction and dip. The components of the tensor oriented toward the same direction of the gradient can contribute to the groundwater flow. In order to evaluate the contribution of the single components of the hydraulic conductivity tensor it is necessary to act as follows:

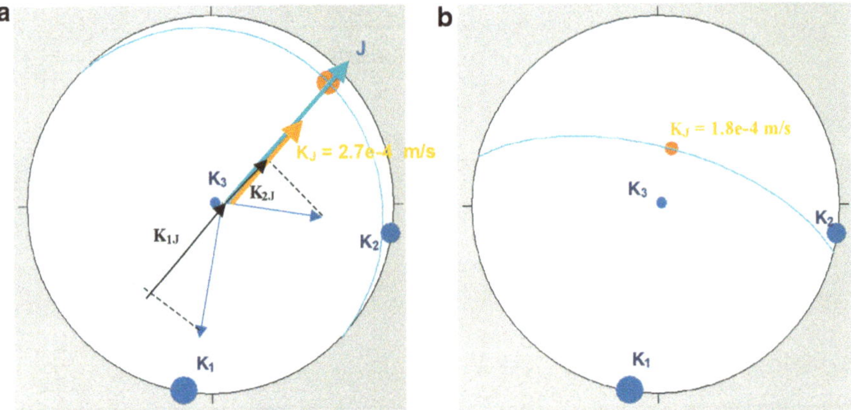

Fig. 4.8 Stereographic representation of hydraulic conductivity (*in orange*) deriving from the composition of the single hydraulic conductivity tensors in the gradient direction: (**a**) With a gradicnt cqual to 45°/10°; (**b**) with a gradicnt cqual to 15°/65°

- to represent both the gradient and the tensor components;
- to project the tensor components in the direction of the hydraulic gradient (on the stereonet it is also possible to create a graphic construction only relating to the horizontal plane);
- to sum the single components to identify the average hydraulic conductivity along the gradient direction.

When the gradient changes, the contribution of the single discontinuity sets to the waterflow changes as well and, as a consequence, also the average hydraulic conductivity along the direction of gradient K_J (whose module is represented in the stereonet by the dimension of the orange circle).

In this way it is also possible to identify the families that contribute to the water flow and, in case, estimate the flow rates.

4.6.2 Unknown Hydraulic Gradient

If the gradient direction is unknown or it is variable according to the water recharge, it is necessary to act as follows:

- to calculate the hydraulic conductivity tensor components;
- to project these components along all the hydraulic gradient orientations (dip direction from 0° to 360°, and dip from 0° to 90°);
- to recombine the above-mentioned components as to find the hydraulic conductivity distribution (apparent hydraulic conductivity) along the different gradient directions (Fig. 4.9).

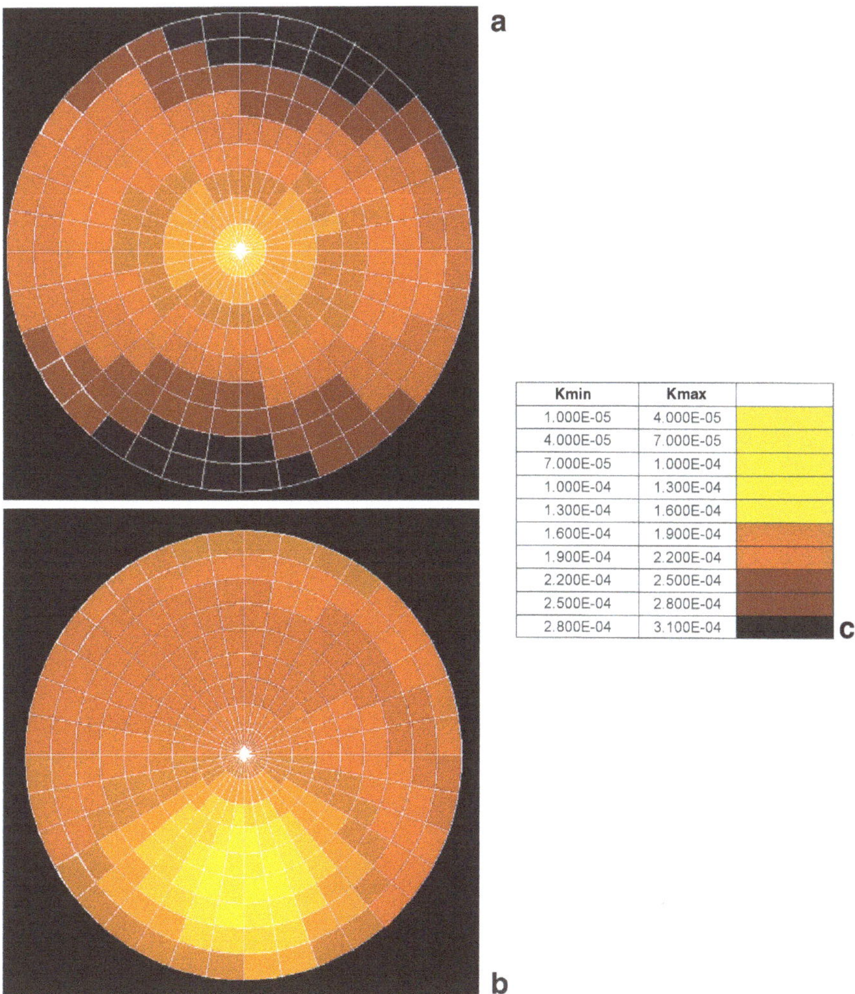

Kmin	Kmax	
1.000E-05	4.000E-05	
4.000E-05	7.000E-05	
7.000E-05	1.000E-04	
1.000E-04	1.300E-04	
1.300E-04	1.600E-04	
1.600E-04	1.900E-04	
1.900E-04	2.200E-04	
2.200E-04	2.500E-04	
2.500E-04	2.800E-04	
2.800E-04	3.100E-04	

Fig. 4.9 Stereographic representation of hydraulic conductivity along the different directions of the hydraulic gradient: (**a**) For the example in Table 4.1; (**b**) with discontinuity sets characterized by the same aperture and frequency and by following orientations: 340°/45°, 30°/30° e 0°/40°; and (**c**) colours scale used in stereographic representations

Chapter 5
Methods and Models to Simulate the Groundwater Flow in Rock Masses

5.1 Introduction

Moving from a qualitative to a quantitative interpretation of the groundwater flow in rock masses, which is fundamental in application problems (for example, to correlate groundwater flow in the rock mass with the development of slope instability phenomena), requires the development of a model of the groundwater flow within the discontinuity network, according to specific boundary conditions. Quite often, the complexity of the problem sets limits to its application. Actually, that approach is quite difficult due to the need to recreate a complex system such as that of the rock mass, where the availability of data in depth is often almost scarce, in particular if quite large areas are taken into consideration.

Different models are available to reconstruct the waterflow in rock masses; some of them refer to an equivalent continuous porous medium (MODFLOW, McDonald and Harbaugh, 1988); others apply a discrete approach that takes into consideration the features of each discontinuity set (FRAC3DVS, Therrien, 1992). A third possibility is represented by dual porosity models, where the flow takes place both in discontinuities and within the porous rock matrix (Shapiro and Andersson, 1983; Rasmussen, 1988).

The choice whether to use the one or the other method is made on the basis of the geological-structural data gathered and the work scale chosen, always considering that the results provided by numerical modeling may contain errors linked both to the limited ability of the model itself to represent the real geological structure and to the difficulty of characterizing the discontinuity network in the scale required.

5.2 Basic Elements of a Modeling Approach

The groundwater flow modeling became quite common in the 1980s and 1990s thanks to the development of hardware and the implementation of new software for mathematical simulation.

Generally, models are very useful tools, as they allow a simplified representation of quite complex natural phenomena. In the case of hydrogeological analysis of an area, the use of a flow model may have three aims:

L. Scesi, P. Gattinoni, *Water Circulation in Rocks*, DOI 10.1007/978-90-481-2417-6_5,

1. *Forecasting aim*, when the aim is the forecast of what happens in a physical system after a change in the boundary conditions.
2. *Interpretative aim*, when the aim of the research is the study of the control parameters of a specified area or a specific phenomenon.
3. *Study of generic geological environments*, considering a possible future analysis of similar real physical systems.

At a generic level, it can be said that the mathematical models simulate phenomena indirectly, through a system of equations. In a flow model, the following equations are used:

- fundamental flow equation;
- boundary conditions (flows along the model boundary);
- initial conditions (initial distribution of the variables, in case of unsteady phenomena).

Once the mathematical model has been created, it can be solved both analytically or numerically, according to the starting hypothesis and the target. Therefore, the model is a very useful and versatile tool in terms of flexibility in approaching the problems, mainly for the possibility to manage more parameters at the same time. Next to these qualities, the use of a model also presents unavoidable disadvantages that must be taken into account: First of all are the costs, due to the bulk of the required incoming data, but its mainly the reliability of results, which are affected by a mistake that increases exponentially if the data are scarcer at the start. Then, it is important to decide what kind of results are expected from the model, as this choice determines the complexity of the model itself and, therefore, the efforts the implementation requires.

The procedure that is generally used when creating a mathematical model can be outlined as follows (Fig. 5.1):

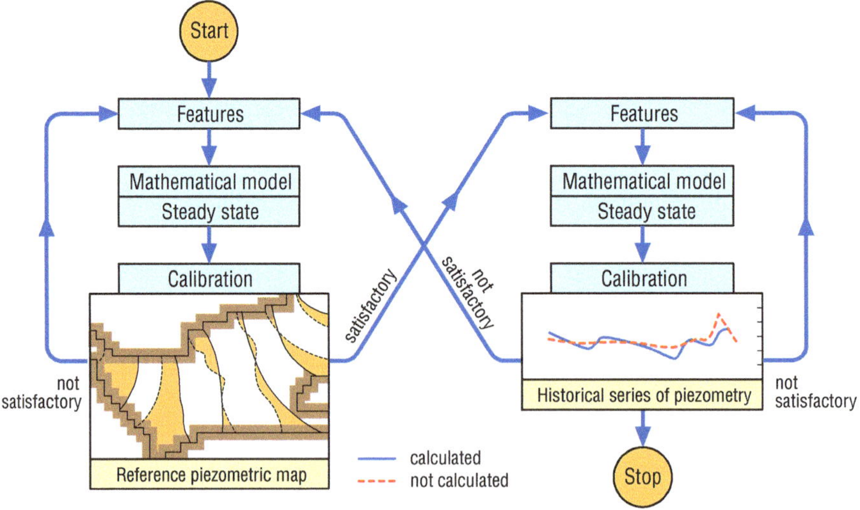

Fig. 5.1 Schematization of the model calibration (Civita, 2005)

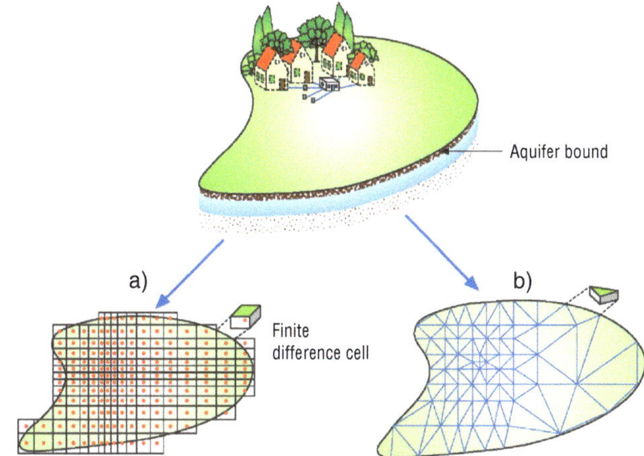

Fig. 5.2 Two-dimensional discretization of the model domain by (**a**) finite differences and (**b**) finite elements (Civita, 2005)

1. *Statement of the aim of the model*: It defines the complexity degree of the model.
2. *Definition of the conceptual model*: It represents reality in a simplified way, defining the model boundaries and the hydrostratigraphic units.
3. *Choice of the general equation and numerical code*: This choice has to be taken according to the kind of problems that must be analyzed, the aims of the model and the informatics tool used.
4. *Model implementation*: It includes the definition of the initial and boundary conditions, the choice of the grid (Fig. 5.2) and the assignment of the preliminary values to the hydrogeological parameters of the materials involved.
5. *Parameters calibration*: It is a first check of the model quality, aimed at testing its ability to reproduce real phenomena.
6. *Sensitivity analysis*: It evaluates the effect of the uncertainty in the estimate of some parameters on the simulation results.
7. Forecasts and interpretation of the results.
8. Possible adjustments of the model.

The last generation of models is represented by stochastic models, based on probabilistic-statistical tools that can analyze the behaviour of a system in a non-deterministic way.

5.2.1 Definition of the Conceptual Model

The *conceptual model* is a schematization of the physical system aimed at idealizing and simplifying the real situation. That simplification is necessary as a complete reconstruction of the real domain is not obtainable due to the excessive pieces of information it would require. On the other hand, it is also true that the closer to reality the conceptual model is, the more accurate the resulting numerical model will be.

The problem in the definition of the conceptual model lies in the need to reconcile contrasting requirements: The conceptual model must be as simplified as possible, according to the complexity of the structure under examination and to the type and quality of the desired result.

Once the aim has been set, the following step is the delimitation of the domain model examined and the identification of its boundary conditions.

5.2.2 The Model Project

The project of the model requires the following steps.

Analysis of the hydrogeological data available for the construction of the conceptual model in the area to be studied

- rainfalls and temperature data (recharge and evapotranspiration);
- stratigraphies of wells and piezometers;
- permeability, transmissivity and porosity of the hydrogeological units identified;
- piezometries and relevant distribution charts of the hydraulic heads within the domain model to be studied and in nearby areas;
- if required, water samples from active wells in the area being studied;
- position and discharge of the springs;
- discharge and water levels of the rivers.

Implementation of the flow model using the numerical code chosen:

- creation of the conceptual model of the system to be studied, with delimitation of the domain model, identification of the geological and hydrogeological target units, reconstruction of the groundwater conditions, assessment of the recharge and groundwater discharge;
- choice of the discretization grid of the domain model;
- definition of the boundary conditions and of initial conditions, according to the information available.

Model calibration:

- calibration of the parameters, in particular in relation to the conductance value to be attributed to streams and springs;
- analysis of the model sensitivity, in relation to the completeness and reliability of the data used.

Simulation and analysis of the scenarios:

- simulation of the groundwater flow conditions with different rainfalls and/or boundary conditions.

5.2.3 Choice of the Numerical Code

Before starting the real project of the mathematical model, the equation ruling the physical process to be analyzed must be identified and the most suitable numerical code for the modeling has to be chosen. First of all, the type of model has to be chosen according to the hydrogeological context and the analysis scale:

1. *Equivalent continuum* (i.g. MODFLOW, McDonald and Harbaugh, 1988).
2. *Discrete approach* that can take into account the features of each single discontinuity (FRAC3DVS, Therrien, 1992).
3. *Dual porosity* with flow both in discontinuities and within the porous matrix (Shapiro and Andersson, 1983; Rasmussen, 1988).

In any case, the equation constituting the flow model can be solved analytically or by numerical approach. Analytical models presuppose simplifications (homogeneity and mono- or two-dimensional flow). For these reasons, solutions using the numerical approach are more common in practical applications. Actually, thanks to the calculation power of modern computers, numerical models are more versatile and allow to carry out a higher number of simulations in relatively short times. The most common numerical methods used to solve groundwater flow problems are the finite elements and finite difference methods; the choice between these two techniques depends on the system to be modeled.

5.3 Darcy's Model

Many authors decided to use Darcy's models also for the schematization of fracture systems, making references to the concepts of equivalent hydraulic conductivity and representative elementary volume. Evidently, that is possible only when a URV (see Chapter 1) an be set within the domain to be studied.

Generally, the continuum choice occurs according to the knowledge on the distribution of fractures and hydraulic conductivity and according to the computational time required by the elaboration. If the waterflow is ruled by a single discontinuity system, an equivalent hydraulic conductivity can be estimated for the rock mass using the hypothesis of the continuum, or a plane-parallel approximation (Domenici and Schwartz, 1990; Hsieh, 2002). On the other hand, hydraulic conductivity can vary in space and the rock mass can be anisotropous; moreover, the presence of extended shear zones highly conditions the groundwater flow. Often, the best solution consists of using a relatively simple model such as that of the continuum, keeping into account the presence of structural features such as shear zones that represent basic elements in the ruling of waterflow. With reference to an equivalent porous medium, some trials were carried out also to determine the transmissivity of karst structures using *back analysis* procedures (Laroque et al., 1999). Generally, the results obtained from the application of Darcy's models underline how strongly the model calibration is influenced by the dimension of the cells used and the availability of field measures.

Models based on the use of the equivalent continuum solve the *fundamental equation of flow in porous media*:

$$K_x \frac{\delta^2 h}{\delta x^2} + K_y \frac{\delta^2 h}{\delta y^2} + K_z \frac{\delta^2 h}{\delta z^2} = \frac{1}{1+n}\left(n\frac{\delta S}{\delta t} + S\frac{\delta n}{\delta t} \right),$$

where

- K_x, K_y and K_z = hydraulic conductivity along the direction x, y and z, respectively;
- h = hydraulic head;
- S and n = saturation degree and porosity, respectively.

Usually, that equation is solved using finite difference and cell-centered methods. In this way, as N is the number of sites, a system of N equations in N unknown (the

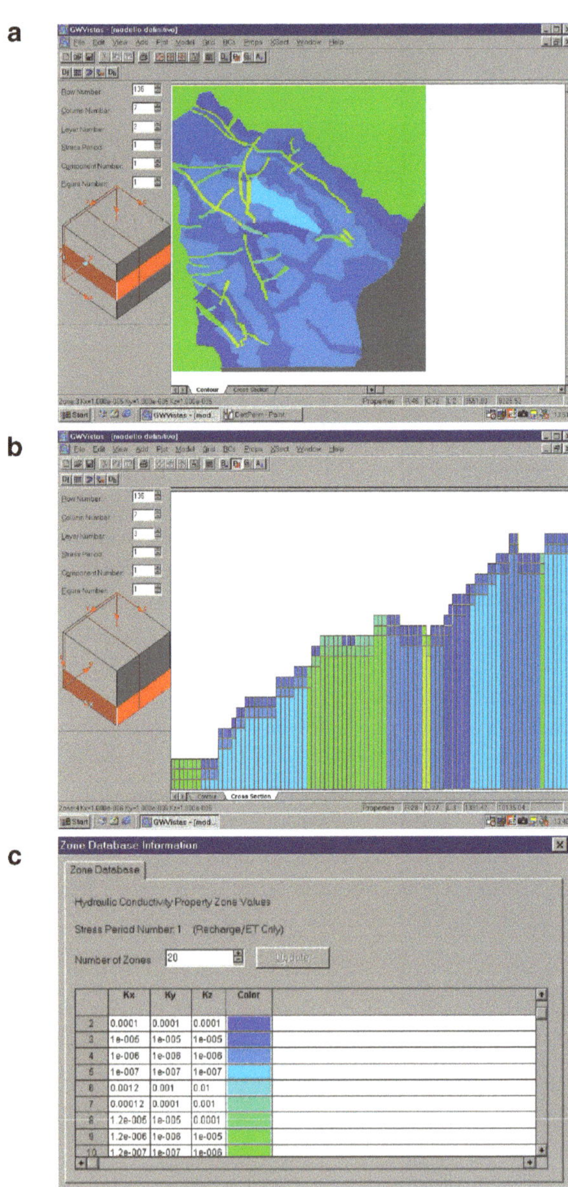

Fig. 5.3 Hydraulic conductivity distribution in the domain studied in planimetry (**a**) and in cross section (**b**: detail) and relative caption (**c**). Take notice of the extreme heterogeneity (both in planimetry and in depth) that characterizes the area and the anisotropy of hydraulic conductivity values, assessed in function of the orientation of the main discontinuity families. The darker areas are the more permeable ones; exceptions are represented by discontinuities, which are presented as green or yellow areas, according to the main flow direction (Croci et al., 2003)

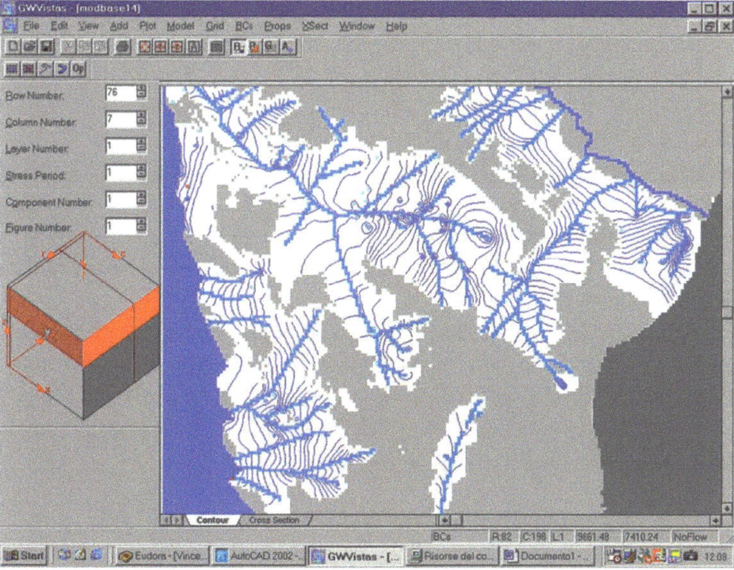

Fig. 5.4 Piezometry obtained from the model. The figure also highlights the boundary conditions imposed on the model: constant head are *blue navy*, rivers are *blue*, no flow cells are *grey*, springs are *cyan*, and wells are *red* (Croci et al., 2003)

piezometric heads) is obtained; the system is solved in an iterative way setting an appropriate convergence criteria. The value of the hydraulic head in the middle of the cell can be calculated in the whole domain model considered.

For each cell in which the domain model is divided, the values of the hydrogeological data (hydraulic conductivity, thickness of the aquifer, storativity coefficient, etc.; Fig. 5.3) are provided. As a fractured medium has to be modeled, in order to attribute the hydraulic conductivity values it is necessary to calculate beforehand the hydraulic conductivity tensor from which the values K_x, K_y and K_z will be obtained to be inserted in the model.

Also boundary conditions have to be included to have a determinate problem. These conditions can be of different kinds (Fig. 5.4):

- Dirichlet conditions (with constant hydraulic head);
- Neumann conditions (with constant flow; i.g. for wells or watersheds);
- Cauchy conditions (with hydraulic head depending on the flow; i.g. for rivers and springs).

5.4 Discrete Models

It is evident that this type of Darcy's approach is no longer valid if the system is studied on a detailed scale, as this scale requires discrete flow models. This implies the need to generate a discontinuity network (see Section 1.6) and to apply the appropriate boundary conditions to it, in order to decide which are the most suitable numerical techniques for the flow forecast.

Discrete models developed both in two dimensions (Long and Witherspoon, 1985; Robinson, 1982) and in a three-dimensional field (Hung and Evans, 1985; Rasmussen, 1988; Andersson and Dverstorp, 1987; Dverstorp and Andersson, 1989), explicitly simulate the flow of each single discontinuity using, for example, the Navier-Stokes equation (Bear, 1993), Kirchoff's laws for electric circuits (Kraemer and Haitjema, 1989) or the model with hydraulically connected circular disks (Cacas et al., 1990).

The simplest *two-dimensional model* of the flow network is the one that assumes the presence of a series of interconnected bonds. The value of the hydraulic head of each site can be calculated as a weighted average of the hydraulic head of the nearby sites, each one multiplied by a transmissivity coefficient N_{ij}, related to the flow between the site i and the site j (Fig. 5.5):

$$H_0 = (N_{10}H_1 + N_{20}H_2 + N_{30}H_3 + N_{40}H_4)/(\Delta N).$$

If transmissivity T_1 is equal to T_2

$$N_{12} = T_1 = T_2$$

If transmissivity T_1 is infinite

$$N_{12} = 2T_1$$

If the two transmissivity values are different, N_{12} is the harmonic mean of the two:

$$N_{12} = (2T_1T_2)/(T_1 + T_2)$$

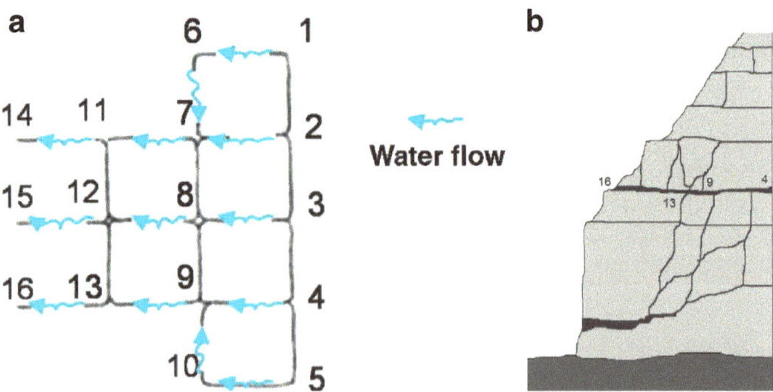

Fig. 5.5 Schematization on a plan (**a**) and in cross-section (**b**) of the two-dimensional discrete flow model (Croci et al., 2003)

Actually, also in the international literature, the application of models based on the flow hypothesis along discontinuities (Diodato, 1994) is either limited to simplified or limited cases (McCaffrey and Adinolfi, 2003; Jeong et al., 1999; Machado et al., 2001) or it is referred to very specific and detailed studies (Oxtobee and Novakowski, 2003); in such studies, nevertheless, the flow conditions are analyzed according to single discontinuities and their parameters, whereas the effects produced by the presence of real shear zones are usually neglected. In that sector, the application of numerical models with distinct elements arises particular interest; they can analyze the groundwater flow both in a two-dimensional and a three-dimensional field. But such an approach requires the availability of rich quantities of data about the underground and, as a consequence, it was used more often for the study of tunnel inflows (Papini et al., 1994; Molinero et al., 2002).

An example of a mathematical model that can simulate the waterflow within discontinuities, using a numerical methods with distinct elements, is the *UDEC calculation code* (Itasca, 1999). Thanks to it, a mechanical-hydraulic study can be carried out, where the joint hydraulic conductivity also depends on the mechanical deformation that, in turn, is influenced by the water pressure inside the discontinuities (Fig. 5.6). That approach allows the reconstruction of the flow network inside the rock masses, thus determining both discharge and water pressures in single discontinuities (Fig. 5.7), and it is particularly useful in the solution of problems regarding slope stability (Fig. 5.8) and the drawing of projects for underground works (Fig. 5.9).

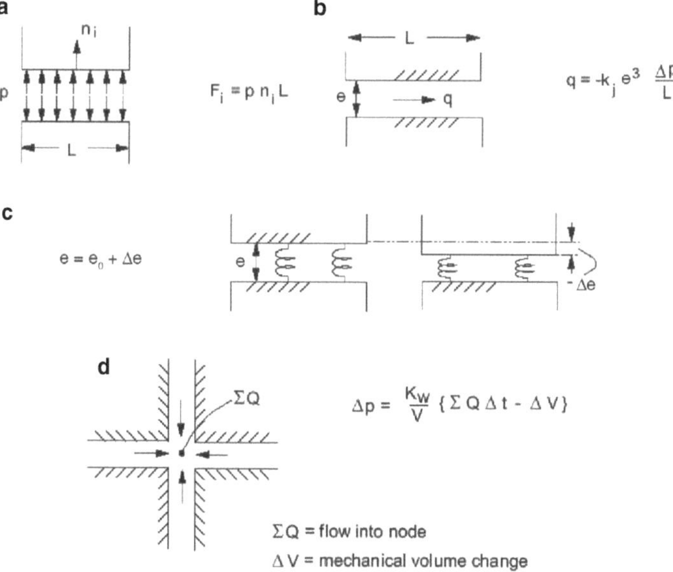

Fig. 5.6 Schematization of the solid–fluid interaction within discontinuities: (**a**) Water pressure; (**b**) water discharge; (**c**) mechanical effects induced by stresses affecting the aperture; and (**d**) pressure change in the sites (Itasca, 1999)

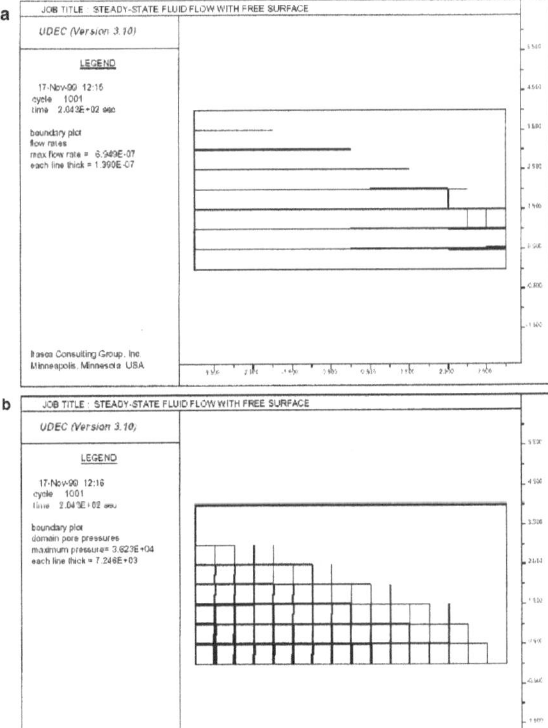

Fig. 5.7 Example of flow modeling with distinct element code UDEC: (**a**) Discharge within discontinuities and (**b**) values of water pressures within discontinuities (Itasca, 1999)

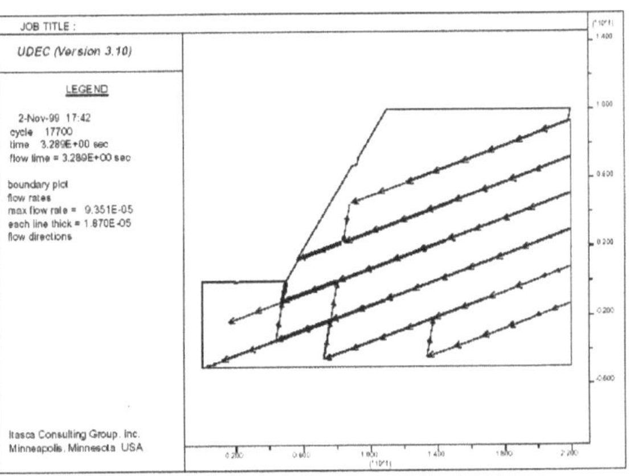

Fig. 5.8 Example of flow modeling in a rocky slope using the UDEC numerical code (Itasca, 1999)

In order to provide satisfactory results, the discrete approach must include detailed information on the characteristics of the discontinuities. Therefore, it allows the simulation of local phenomena or detailed scale processes, whereas it is not suitable to model generic scale processes, due to the computational limitation linked to the large quantities of required data. Often, the discrete approach is appropriate

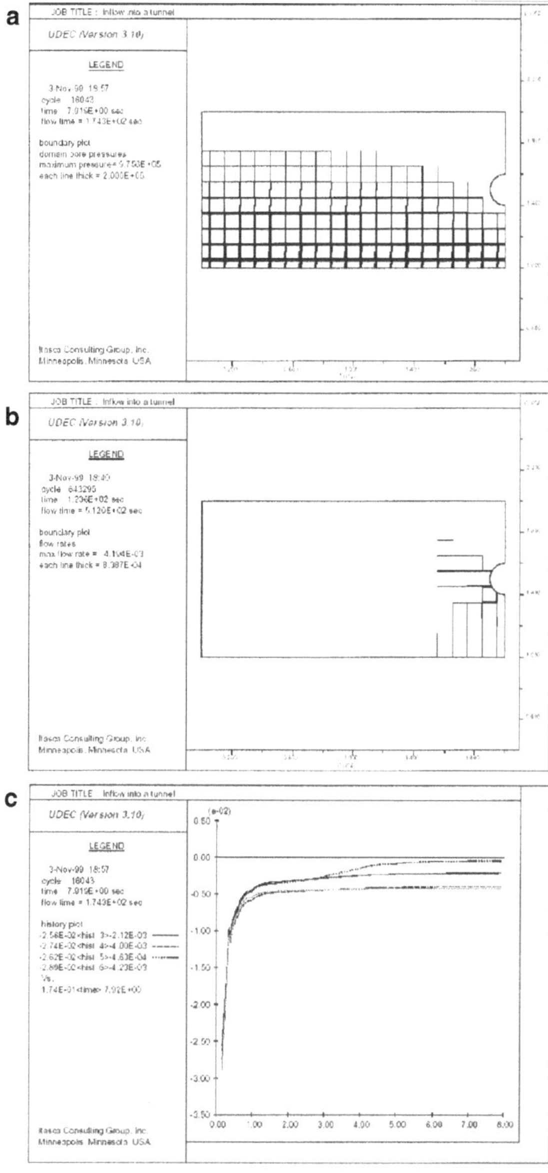

Fig. 5.9 Example of modeling of the draining process induced by a tunnel using the UDEC numerical code: (**a**) Trend of water pressures within discontinuities; (**b**) water discharges within discontinuities; and (**c**) trend in time of the tunnel inflows (Itasca, 1999)

when there are discontinuity systems of localized shear zones that are ruling the water flow (Samardzioska and Popov, 2005).

5.5 Dual Porosity Models

The discrete approach is originated from the observation that non-fractured rocks may present high porosity (storage coefficient) with very low hydraulic conductivity (lack of flow); on the contrary, high hydraulic conductivity (and therefore flows) can be present within fractures, even with small storage volumes.

Dual porosity models (introduced by Barenblatt et al. in 1960 and then implemented by Warren and Root in 1963 and Kazemi in 1969) try to combine the simplification of Darcy's models with the complexity of discrete models, taking into account, at the same time but still separating from one another, the water flow within intact rock (*primary porosity*) and that of the discontinuity network (*secondary porosity*).

Actually, in dual porosity models, equations are used that can rule both continuous and fractured media, among which flow exchange at the interface is possible (Fig. 5.10). Some examples of semi-analytical and numerical models were proposed by Huyakorn et al. (1983), Rowe and Booker (1990) and Sudicky (1990). Generally, dual porosity models are used to simulate the flow and transportation of contaminants (Bai et al., 1997; Moutsopoulos et al., 2001; Alboin et al., 2002) as these elements may have relevant interaction with the rock matrix, mainly in relation to the propagation and dispersion of the contaminant. Some limitations of dual porosity models are represented by the little realistic adoption of a very simplified geometry of rock masses and by the fact that the advection in the rock matrix is generally neglected.

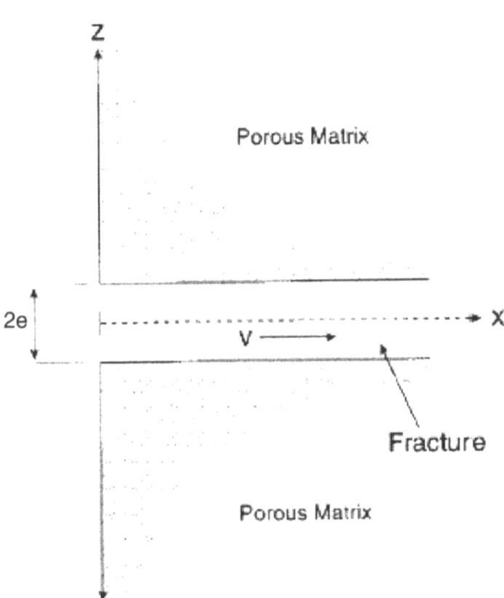

Fig. 5.10 Schematization of a fracture in a porous medium (Delleur, 1999)

Example 1

With reference to Example 1 in Chapter 1, a discrete flow model was created by 3D UDEC.
The schematization of the domain model obtained through the discontinuity generation is presented in following figures.

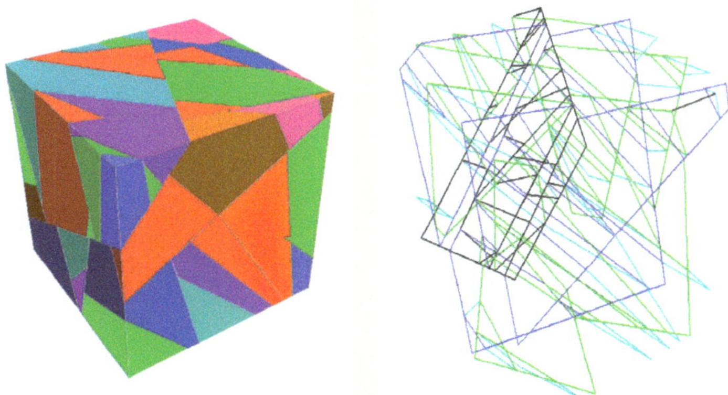

Then, a constant head condition was applied (in the left) and the flow was simulated.

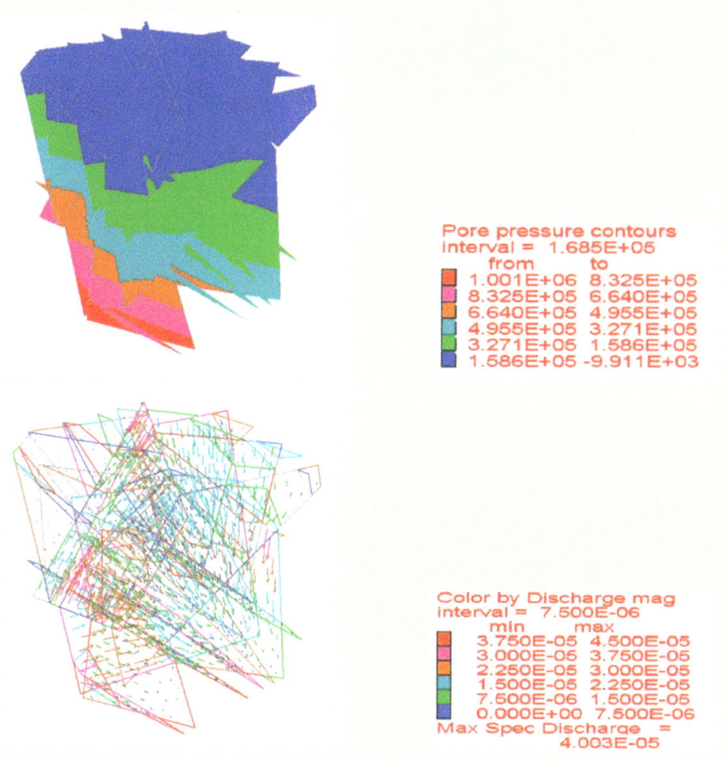

Chapter 6
Case Histories

6.1 Groundwater Flow and Slope Stability

The first example regards the study carried out on the Roccaprebalza serpentine relief (Val Taro, municipality of Berceto, in the province of Parma, Northern Italy; Fig. 6.1). Here groundwater flow was analyzed in order to find possible correlations with the slope mass movements of this area, situated at the border of the large and deep gravitational deformation affecting Berceto.

From the lithologic point of view, the outcropping rocks are ophiolites resting on a clayey bedrock belonging to the Clay Blocks Complex (Fig. 6.2). From the geological–structural point of view, the area is characterized by the presence of an important East–West fault, that is part of a Northeast–Southwest fault system. The high average hydraulic conductivity of the rock masses was estimated through the Lefranc tests in values ranging from 10^{-3} to 10^{-8} m/s, and it is due to the fracturing degree of the outcropping rock in the Roccaprebalza area.

Fig. 6.1 Picture of Roccaprebalza (Berceto, Northern Italy)

L. Scesi, P. Gattinoni, *Water Circulation in Rocks*, DOI 10.1007/978-90-481-2417-6_6,

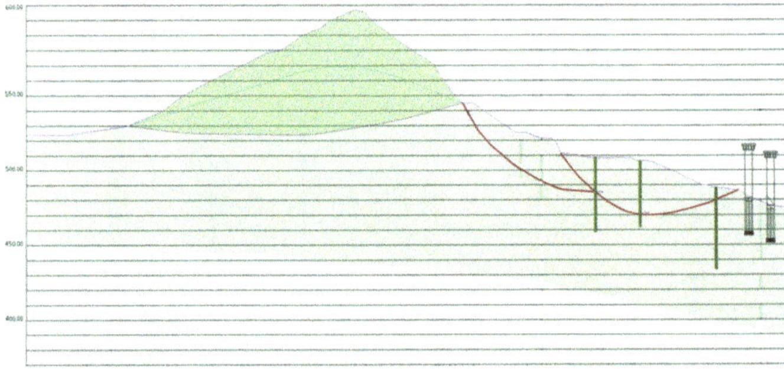

Fig. 6.2 East–West section of Roccaprebalza. Shale is represented in *grey*, serpentinite is represented in *green*

Fig. 6.3 Picture of the outcropping where the geostructural survey in the example of Fig. 6.4 was carried out

The first step to recreate the hydrogeological setting of Roccaprebalza was a geological–technical and structural survey on some outcroppings covering a surface of about 40 m^2 (Fig. 6.3), aimed at obtaining the typical parameters of the discontinuity in terms of orientation and hydraulic features (Fig. 6.4 and Table 6.1). Then, these data allowed to determine the hydraulic conductivity tensor (Table 6.2 and Fig. 6.5a) and identify the main flow direction

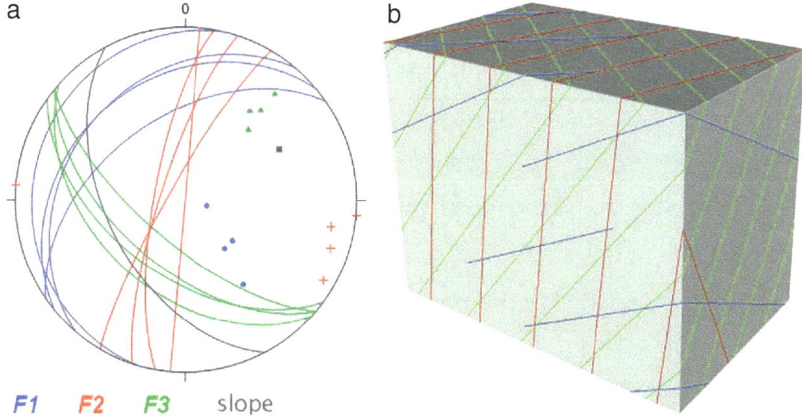

Fig. 6.4 Structural setting of the rock mass (RGM of Fig. 6.1): (**a**) Stereographic projection of the discontinuities that can be grouped in three families; (**b**) tridimensional view of the discontinuity families, where their persistence and spacing is clearly visible

Table 6.1 Orientation and hydraulic features of the discontinuity families identified in RGM1

a)		Average aperture (m)	Average frequency (1/m)	Average spacing (m)	Hydraulic conductivity modulus (m/s)
		e	*f*	*s*	K_i
F1	Dip direction 316° Dip 30°	0.001	0.714	1.4	5.837×10^{-4}
F2	Dip direction 285° Dip 81°	0.0025	0.625	1.6	7.980×10^{-3}
F3	Dip direction 220° Dip 57°	0.001	0.833	1.2	6.810×10^{-4}

b)	Length (m)	Normalized interconnectivity index	Angle between the two planes (°)
	l	*I*	γ
	5	0.287	52.413
	15	0.396	30.191
	10	0.317	68.282

- in the non-saturated zone (Fig. 6.5b and Table 6.3),
- in presence of an impermeable bedrock having orientation 50°/20° (Fig. 6.6a),
- in the saturated zone as a function of the hydraulic gradient changes (Fig. 6.6b).

Table 6.2 Hydraulic conductivity tensor and equivalent hydraulic conductivity of the rock mass in RGM1

	(m/s)	Dip direction (°)	Dip (°)	K_{eq} (m/s)
K_1	9.12×10^{-3}	243	9	
K_2	8.84×10^{-3}	154	7	3.49×10^{-3}
K_3	5.27×10^{-4}	100	79	

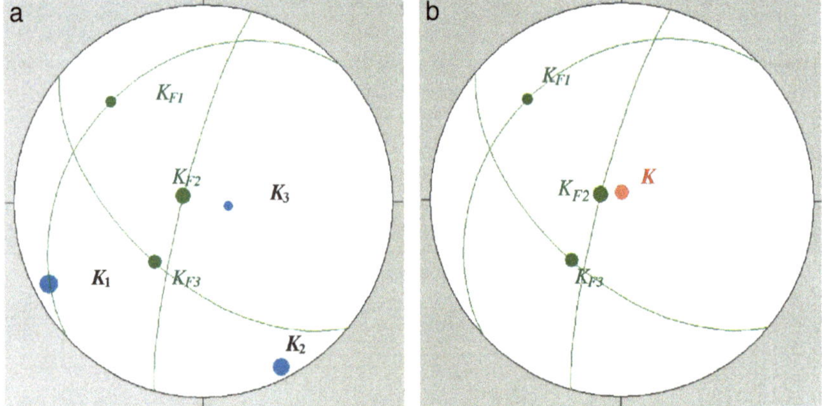

Fig. 6.5 Stereographic representation of (**a**) the hydraulic conductivity tensor in the rock mass (*in blue*); (**b**) of single hydraulic conductivity oriented along the discontinuity planes (*in green*) and of the main flow direction in a non-saturated medium (*in red*)

Table 6.3 Resultant of the composition of vectors relating to the percolation along the different discontinuity families

Vectorial sum K_{tot}		
	x	3.377×10^{-5}
Vector comp.	y	4.105×10^{-4}
	z	-7.517×10^{-3}
Modulus (m/s)	K_{tot}	7.528×10^{-3}
Dip direction	$\alpha\ (K_{tot})$	4.70°
Dip	$\psi\ (K_{tot})$	86.86°

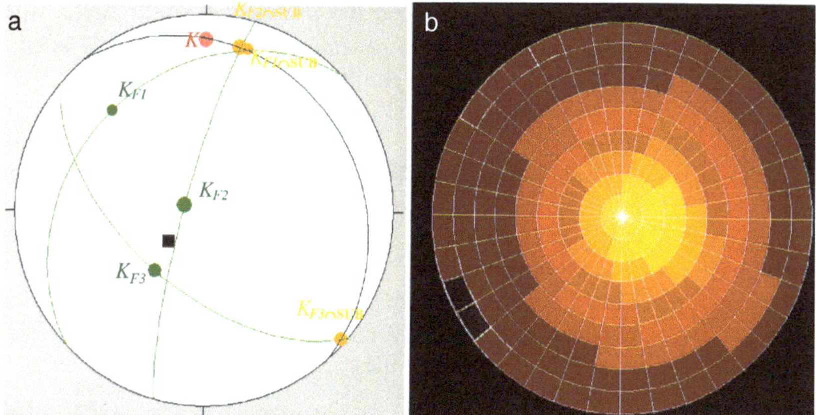

Fig. 6.6 Stereographic representation of (**a**) the main flow direction with an impermeable bedrock; (**b**) of hydraulic conductivity in the different directions of the hydraulic gradient (the color scale has a 10^{-3} m/s interval)

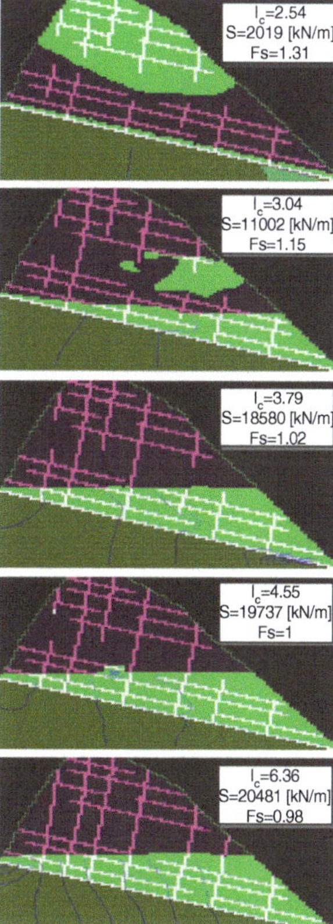

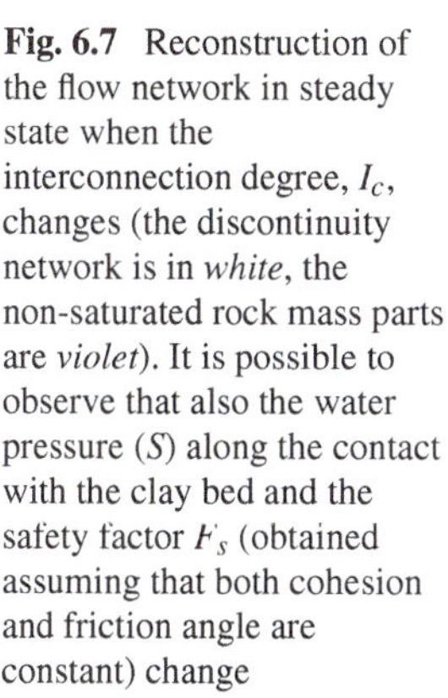

Fig. 6.7 Reconstruction of the flow network in steady state when the interconnection degree, I_c, changes (the discontinuity network is in *white*, the non-saturated rock mass parts are *violet*). It is possible to observe that also the water pressure (*S*) along the contact with the clay bed and the safety factor F_s (obtained assuming that both cohesion and friction angle are constant) change

In this way it was possible to recreate the groundwater flow paths within the Roccaprebalza, to identify the recharge and discharge areas and to relate them with the stability conditions of the slopes that are strongly affected by the depth of the water paths and therefore by the interconnectivity among different discontinuity sets.

In order to assess quantitatively the effect of interconnectivity on the stability conditions of the slope adjoining Roccaprebalza, a flow model with finite differences was implemented, where the fractures were simulated as highly permeable cells. The simulation was carried out with different values of the interconnectivity index that allowed to evaluate the water pressure when getting in contact with the clay bedrock and the consequent sliding safety factor (Fig. 6.7). This study highlighted that when the interconnectivity increases, the groundwater discharge reaching the

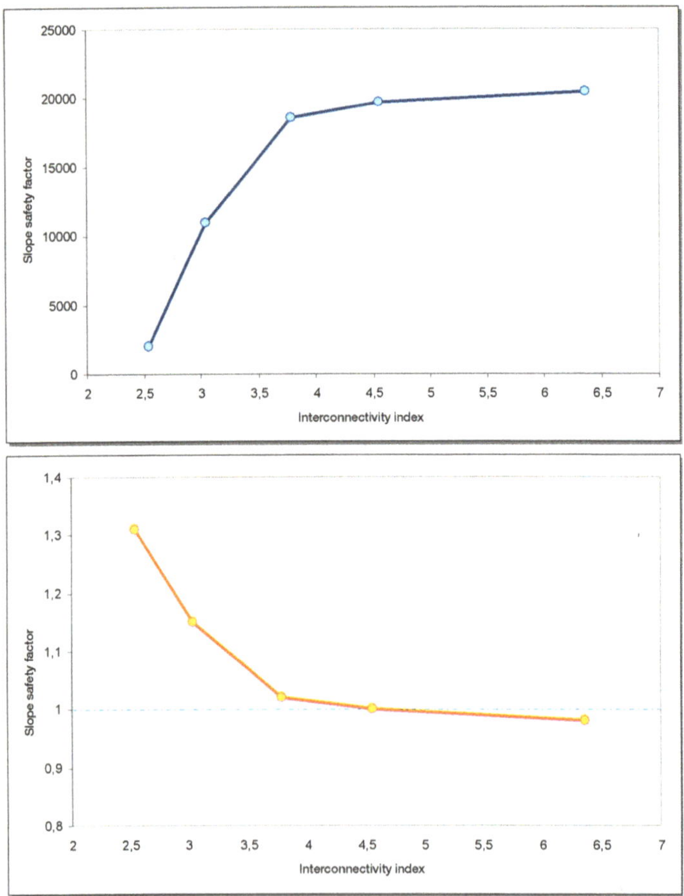

Fig. 6.8 Trend of the water pressure along the contact area with the clay layer and of the safety factor to sliding along the same area when the interconnectivity index changes

bedrock also increases and, as a consequence, the water pressure along the contact; if the resistance values of the material are the same, when the interconnectivity degree increases, this leads to a reduction of the safety factor (Fig. 6.8). Actually, when the interconnectivity degree is low, water circulation is mainly superficial and the possible slope instabilities that may be triggered are represented by sliding and/or toppling of single blocks, as observed along the Southern slope of Roccaprebalza

Fig. 6.9 Picture of Roccaprebalza: (**a**) south slope, with a widespread groundwater flow and deep instability; (**b**) north slope, with localized groundwater flow and superficial instability

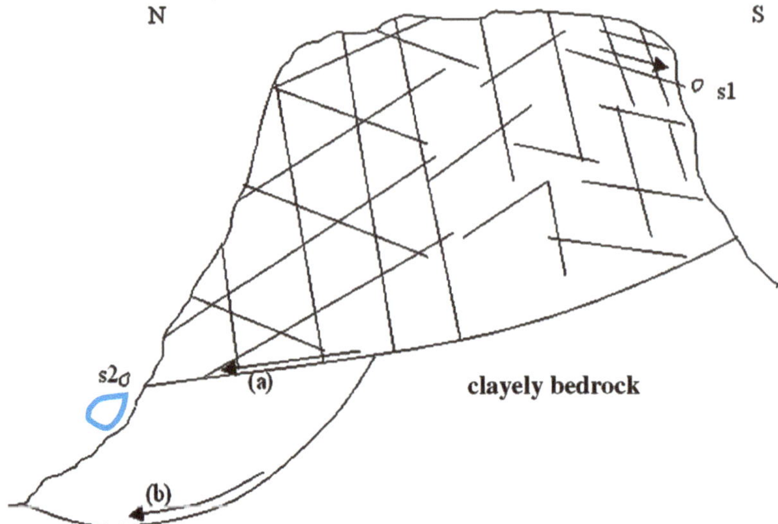

Fig. 6.10 Exemplifying scheme of the interconnectivity degree of Roccaprebalza discontinuities. On the southern slope a structural setting can be detected, which is characterized by a partial interconnectivity among discontinuities, with localized water flow and triggering of superficial instability phenomena. On the north slope, the structural setting is characterized by total interconnection among discontinuities, with water circulation that reaches the bedrock, and triggering of instability in depth: (a) planar sliding along the clay layer; (b) rotational sliding involving also the bedrock. This situation is typical for the north slope of Roccaprebalza

(Fig. 6.9). On the contrary, if the interconnection degree is high, water percolation paths may lead to the creation of groundwater along the clayey bedrock, thus triggering deep instability phenomena linked to planar sliding along the same layer (Fig. 6.10a) or involving the underlying clay in rotational movements as well (Fig. 6.10b).

6.2 Evaluation of the Hydrogeological Risk Linked with Tunneling

The interaction of tunneling with groundwater has become a very relevant problem not only due to the need to safeguard water resources from impoverishment and from the pollution risk, but also to guarantee the safety of workers and the effectiveness of the draining works. The most emblematic examples in Italy concern the construction of the Gran Sasso motorway tunnels (Cotecchia, 1993), where water inflows exceeded 2000 l/s, and the railway tunnel for the Bologna–Firenze high-speed stretch (Rossi et al., 2001), with drained flows reaching 650 l/s. On the contrary, when the tunnels in San Pellegrino Terme were constructed, the interdisciplinary approach of the study and the attention to the environment guaranteed the safeguard of the hydrothermal springs of the area (Barla, 2000).

It is well known that the excavation of tunnels has a relevant draining effect leading to a more or less generalized drawdown of the groundwater level, whose effects may be undesirable, such as the drying up of springs and/or wells (Gisotti and Pazzagli, 2001), qualitative changes of the groundwater (Civita et al., 2002), changes in the vegetations, changes in the slope stability (Picarelli et al., 2002), changes in the flow and quality of thermal waters and changes of the hydrogeological balance at the basin scale. At a general level, it can be stated that the effect of a tunnel on the hydrogeological setting depends on the feeding conditions and on the aquifer permeability, as well as on the system used to excavate the tunnel (Reuter et al., 2000). In the last few years, many studies have been carried out that allowed to define more clearly the contribution that hydrogeology can provide to the different stages of tunnel projecting (Civita et al., 2002), in particular in relation to the problem of forecasting tunnel inflows (Goodman et al., 1965; Knutsson et al., 1996; Ribacchi et al., 2002; Anagnostou, 1995; Federico, 1984; Karlsrud, 2001; Loew, 2002; Molinero et al., 2002; Dunning et al., 2004) and to the impact on the hydrogeological conditions of the surrounding environment, in particular on the regime of springs, groundwater and superficial waters (Dematteis et al., 2001).

From the environmental point of view, the impoverishment, the drying up and, more in general, the change of regime of springs are only some of the most difficult risks to forecast and quantify when projecting a tunnel, as these elements are ruled by very complex and often unpredictable phenomena. Moreover, considering that the hydraulic characteristics of the rock mass are neither homogeneous nor isotropic, the water flow is ruled by the orientation and the hydraulic characteristics of the joints, as well as by the fracturing degree. In shear zones, for example, hydraulic conductivity increases (by some orders of magnitude) can occur along strikes predetermined by the orientation of the same shear zones; the orientation of that flow strongly influences the form and the extension of the area that is potentially interested by the draining process.

The examples presented analyze the main mechanisms ruling inflows in middle-depth tunnels excavated in fractured rock masses, defining the meaning and the dimensions of the areas affected by the works both in saturated and non-saturated conditions (percolation), according to the geological–structural and the hydrogeological setting of the area, with a specific reference to the interconnection degree among discontinuities and among the different aquifers. The case history is used to present a methodology of study that includes the use of both methods of geognostic investigations and statistical analyses aimed at the assessment of the hydrogeological risk in tunneling. In particular, the case of a small-diameter tunnel is analyzed: the tunnel was not impermeabilized and was excavated at medium depth in prevailing flyschoid rocks (Lombard Prealps, Northern Italy) characterized by a middle-low hydraulic conductivity; the realization of the tunnel had a negative impact on the regime of some springs. The study is articulated in two main phases

- the definition of the perimeter of the area affected by geological risk; this operation is carried out by reconstructing the groundwater flow in different conditions, assessing the tunnel inflow and the radii of influence;
- the statistic quantification of the hydrogeological risk for both the tunnel and the environment, through the calculation of the probability that the tunnel inflow or the piezometric drawdown due to the excavation exceeds the acceptable values.

The applicative interest of the study is linked to the possibility of carrying out a probabilistic analysis of the geological risk during the planning of underground construction. This would allow those responsible for the project, and later the project manager, to consider the different risky scenarios and therefore to plan all the required measures to prevent their occurrence, as well as to limit their consequences.

6.2.1 Reconstruction of the Groundwater Flow

In order to bound the potentially interested area by the tunnel construction, at the hydrogeological level, it appears extremely important to integrate the geological, geological–structural, geomechanical and traditional hydrogeological studies with more in-depth analyses about the individuation and the characterization of the "shear zones" and karst circuits. The interpretation of the results obtained by these analyses allows a quite detailed reconstruction of the conceptual model of the groundwater flow. In particular, the following steps are fundamental:

- the identification of the tunnel sections situated over the water table, where only the water percolating inside the discontinuities directly intercepted by the excavation are drained, and those sections fully situated in the groundwater, where a general drawdown of the piezometric level is expected (Fig. 6.11);
- identification of possible intercommunication areas among different aquifers (Fig. 6.12).

In presence of different aquifers, it is necessary to identify the possible interconnection areas by means of focused surveys, carried out in the critical tunnel sections. For example, high-definition geoseismic reflection surveys allow to point out the presence and the trend of very fractured rock zones that constitute the supply areas of deep aquifer circuits. Figure 6.13 provides an example of the interpretation of the geoseismic survey results that highlight the presence of some anomalies in the propagation speed of seismic waves. Those anomalies have to be related to the presence of very fractured rock zones that probably constitute the supply areas of deep circuits that can locally connect two or more aquifers. In this case, the draining process induced by the tunnel excavation might lead to consequences on the deeper aquifer directly interested by the excavation as well as on more superficial waters, having potential negative effects on the regime of springs and rivers in the area.

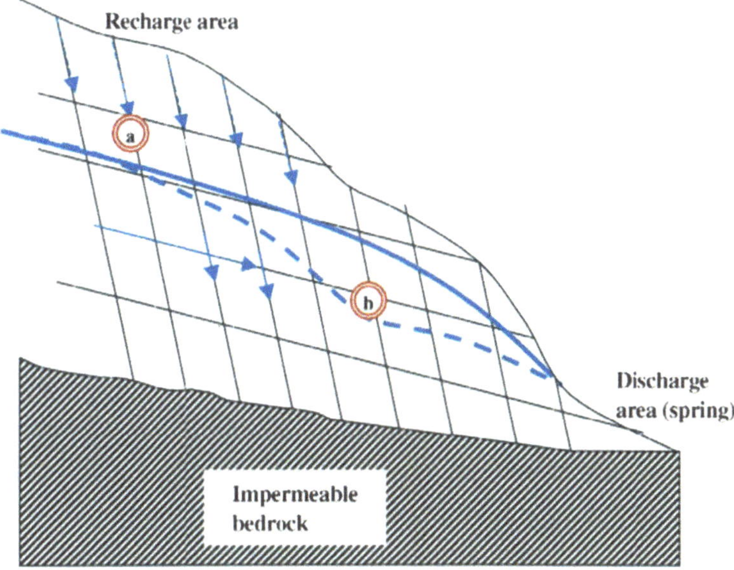

Fig. 6.11 Schematic representation of the flow condition typical of a rock slope, with indication of the water flow direction within the discontinuity network. The water path is conditioned by the hydraulic gradient and by the discontinuity distribution and orientation. The tunnel in position (**a**) only drains water flowing within the discontinuities directly intercepted; the tunnel in position (**b**), i.e. underneath the groundwater, causes a generalized drawdown of the water table, whose shape and extension are ruled by the hydraulic conductivity tensor

Therefore, the presence of shear zones where rock fractures are mainly vertical determines a local interconnection between the deep aquifer intercepted by the tunnel and superficial waters (Fig. 6.14); as a consequence, the superficial catchment area must be defined according to the hydraulic features of the previously identified shear zones.

6.2.2 Estimation of the Tunnel Inflow

An estimate of tunnel inflows can be obtained using geomechanical classifications (Gates, 1997), through analytical formulations (Jacob and Lohman, 1952; Kawecki, 2000; Goodman et al., 1965) or the implementation of mathematic models (Dunning et al., 2004; Molinero et al., 2002), notwithstanding all the mistakes deriving from the uncertainty of the variables involved. Actually, it must be considered that the results obtained are highly conditioned by the hydraulic conductivity value, which depends on the fracturing degree and the stress degree, as well as on the hypothesis of isotropy and homogeneity that constitutes the basis of traditional formulations, which, therefore, are often inadequate to describe correctly the draining process in rock masses.

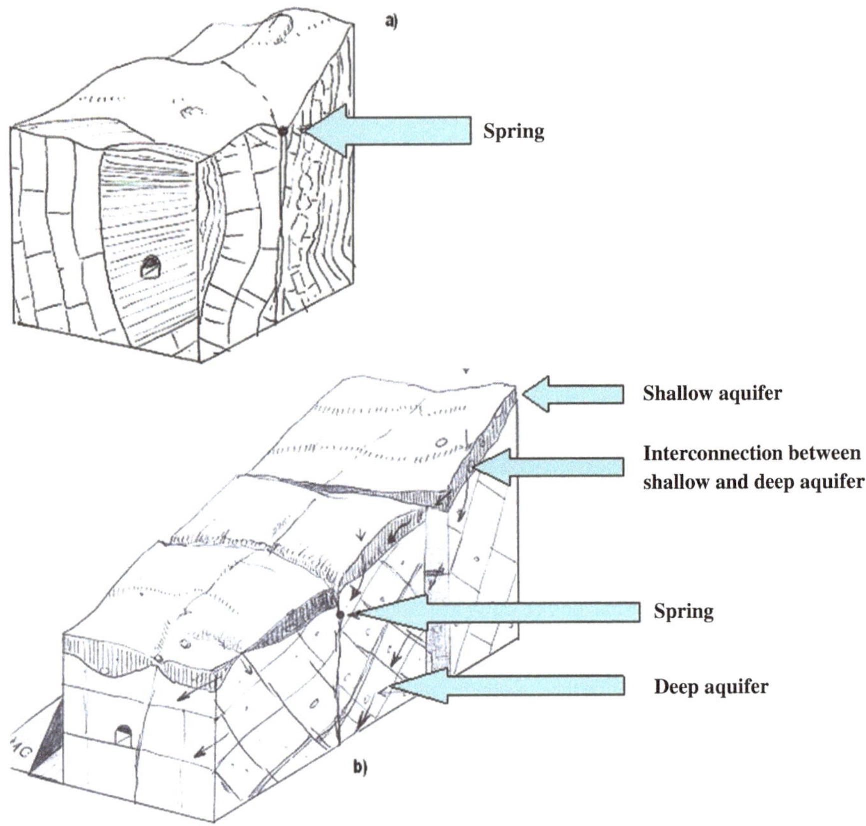

Fig. 6.12 Example of typical schemes of water flow in structures with tunnels at middle-depth: (**a**) the Quaternary overburden (eluvio-colluvial) is quite scarce and the deep groundwater flow is ruled by the fracturing degree of the rock, with water fed by rainfalls, which emerges at the less impermeable layers or peculiar tectonic structures (for example, faults); (**b**) the hydrogeological setting is more complex due to the presence of geological and structural limitations, as well as highly fractured zones that can cause the interconnection among superficial groundwater that feeds the springs and the deeper groundwater intercepted by the tunnel (drawings by Francani)

A semi-quantitative technique to assess if a rock mass can be affected by a high groundwater flow was studied by Gates (1997), who developed a classification of the rock mass called HP (Hydro-Potential or hydrological potential). That classification mainly derives from Barton classification (Q, Barton and Choubey, 1977) and it expresses the HP value as

$$HP = \left[\frac{RQD}{J_n}\right] \cdot \left[\frac{J_r}{J_k \cdot J_{af}}\right] \cdot J_w,$$

where

- RQD, J_n and J_r = parameters used in Barton classification;
- J_k = hydraulic connectivity of the joint;

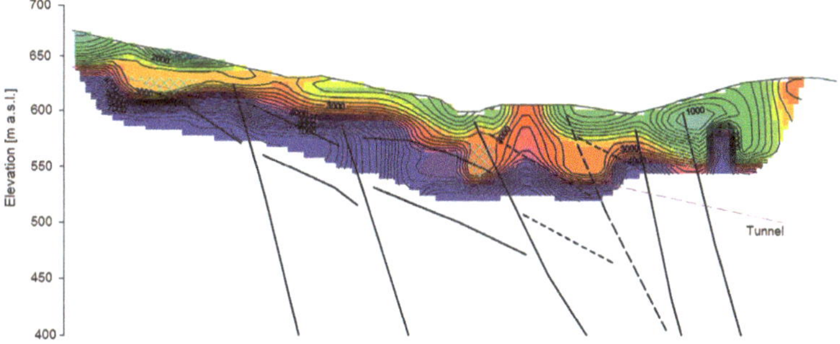

Fig. 6.13 Example of high-definition reflection seismic profile. Speed is expressed in m/s. The *blue dot* formation shows the shear zones, the *red dotted line* shows the tunnel lay-out

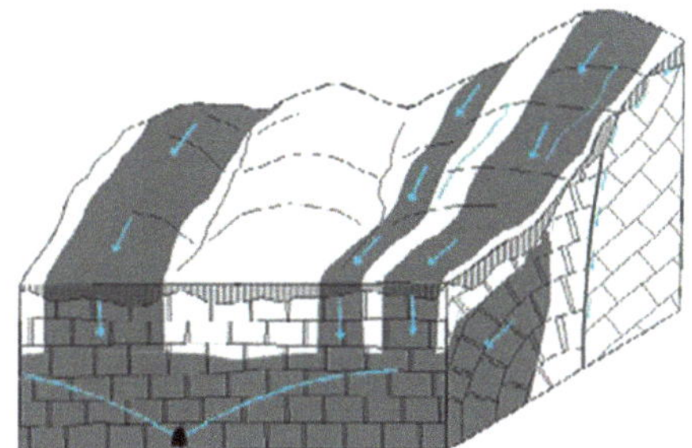

Fig. 6.14 Tridimensional schematization of the shear zones that cause the interconnection between superficial and deep water, with the associated groundwater flow scheme

Table 6.4 Values of the J_w index when rock mass flow conditions change

Conditions	Drops/minute	Liter/minute	J_w
Dry	<1		1
Wet	1–10		0.94
Drops	10–100		0.86
Dripping	> 100		0.76
Diffused seepage		0.0075–0.075	0.66
Low flow		2.3–6	0.5
Average flow		6–60	0.33
High flow		>60	0.2

- J_{af} = joint aperture factor (equal to 1 for closed joints and 2.5 for apertures up to 20 mm);
- J_w = factor that identifies the hydraulic condition at present in the rock mass (Table 6.4).

If the HP value thus obtained is higher than 3, no relevant hydraulic circulation phenomena are forecasted; on the contrary, if HP < 3, the rock mass is potentially interested by hydraulic circulation, whose flow (measured in liters/minute) is estimated as

$$Q = 3.785 \cdot (919.71e^{-2.314\text{HP}})$$

The tunnel inflow can be assessed also starting from the identification of the main direction of the flow.

In a non-saturated rock mass, the water flow inside discontinuities is ruled by gravity, therefore, water tends to flow in depth following the orientation of discontinuities, especially the ones with high hydraulic conductivity (Fig. 6.15). The water percolation goes on till it reaches either a water table (in which case a flowing condition occurs in the saturated medium, ruled by the hydraulic gradient of the aquifer) or an impermeable layer that acts as bedrock.

Therefore, the flow directions during the percolation phase coincide with the dip direction of each discontinuity family. The main flow direction is calculated by combining the discontinuity orientations also considering the above-mentioned flow components along the discontinuity sets (Fig. 6.16). To take into account the interconnection parameter, these components can be considered according to the interconnection degree of each family with the system (Gattinoni et al., 2005).

In the *non-saturated conditions* described above, the tunnel only drains the water percolating within the intercepted discontinuities; as a consequence, the drainage of the tunnel can be estimated, but it is impossible to assess the real radius of influence.

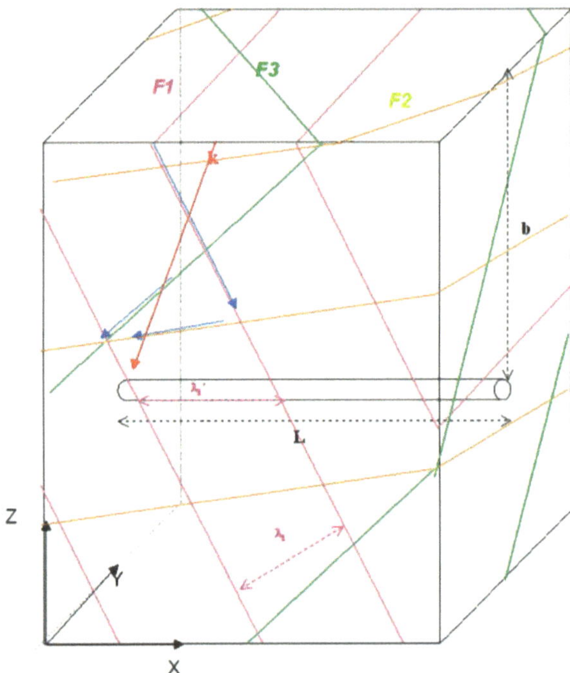

Fig. 6.15 Tridimensional view of the discontinuity sets associated to the three main percolation directions (*in blue*) and of the main flow direction (*in red*)

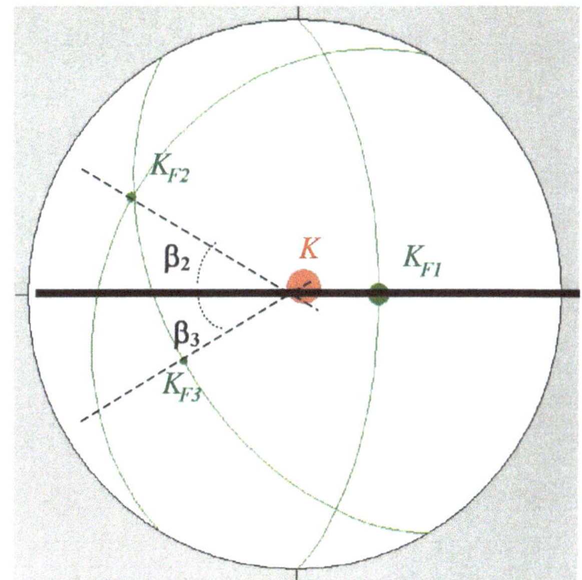

Fig. 6.16 Example or stereographic representation of the main flow direction in a non-saturated medium (*K*, *in orange*). The dimension of *colored circles*, representative of hydraulic conductivity vectors of single discontinuity families (K_{F1}, *in green*), is proportional to the modulus of hydraulic conductivity itself

As far as the flow is concerned, it depends mainly on the probability that the tunnel intercepts the discontinuities belonging to the different sets present; that probability is a function of

- relative orientation of discontinuity and tunnel;
- spacing and persistence of the discontinuities;
- length and diameter of the tunnel.

The spacing of the ith discontinuity family is called λ_i, then the intercept ($\lambda_i\prime$) has to be calculated and, as a consequence, the discontinuity frequency ($f_i\prime$) along the tunnel direction

$$\lambda_i' = \frac{\lambda_i}{sen\alpha_i^{app}\cos\beta_i} \Rightarrow f_i' = \frac{1}{\lambda_i'},$$

where

- $\alpha_i{}^{app}$ = apparent dip of the ith discontinuity family in the tunnel direction,
- β_i = angle between dip direction of the discontinuity and tunnel direction.

If the average length of the trace of the ith discontinuity family, l_i, is known, it is therefore possible to define the intersection probability among discontinuities and the tunnel axis as follows:

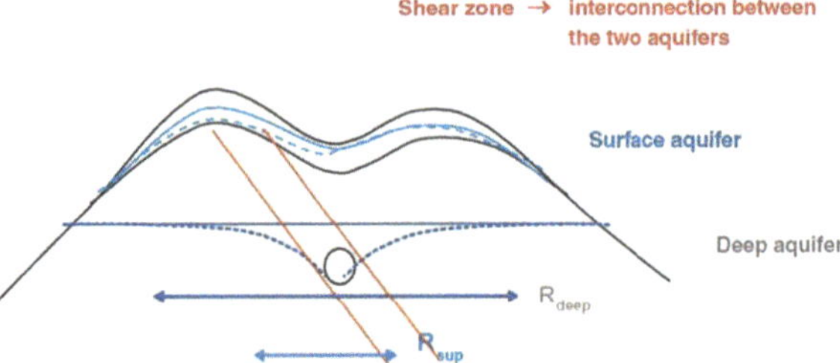

Fig. 6.17 Exemplification of the draining effects of the tunnel on the two aquifer levels (superficial level in *light blue* and deep level in *blue navy*) at a shear zone that causes the percolation in the non-saturated medium along the direction identified by the *red lines*. The *lines* represent the undisturbed piezometric surface, whereas the *dotted lines* represent the new piezometry following the tunnel draining effect on the two water tables

$$p_i = 1 \quad \text{if } I_{i_{tun}} = l_i f_i' > 1,$$
$$p_i = l_i f_i' \quad \text{if } I_{i_{tun}} = l_i f_i' < 1.$$

In a tunnel section having length L, the average number of the intercepted discontinuity belonging to the ith family can be defined as

$$N_i = P_i L = l_i f_i L.$$

The average hydraulic conductivity of the ith discontinuity family is called K_i and e_i is its average aperture; the tunnel inflow is given by

$$Q = \sum_i K_i \frac{2rLsen\alpha_i^{app}}{\lambda_i \cos \beta_i} e_i.$$

In the not saturated area (percolation condition), on the base of the main flow direction, the supply area of the water intercepted by the tunnel can be easily identified. Local drawdown of the piezometric water table in superficial aquifers could occur in those areas (Fig. 6.17).

Concerning the *complete saturated condition*, the literature provides some analytical formulations that allow an approximate estimation of the tunnel inflow (Table 6.5).

It is, nevertheless, useful to remember that in saturated rocks the water flow is ruled by the hydraulic conductivity tensor; therefore, in order to apply the formulas above, the hydraulic conductivity equivalent must be used (Section 2.2.4).

6.2.3 Delimitation of the Tunnel Influence Zone

In the hypothesis of horizontal water table in an isotropic medium, the draining effect of the tunnel can be studied through the above-exposed analytical formulations that allow to implicitly estimate also the corresponding influence zone.

Table 6.5 The analytical formula about the water tunnel inflow. All the above-cited formulas are based on the hypothesis of homogeneous and isotropic aquifer, horizontal water table and $r << H$

Steady state		
$Q = \frac{2\pi kL(H-h)}{\ln\left(\frac{2H-2h}{r}\right)}$	Goodman et al. (1965)	Water table below land surface. Hydrostatic load constant along the tunnel border.
$Q = \frac{2\pi kL(H-h)}{\ln\left(\frac{H-D-h}{r}+\sqrt{\left(\frac{H-D-h}{r}\right)^2-1}\right)}$	Lei (1999)	Water table above land surface. Hydrostatic load constant along the tunnel border.
$Q = \frac{2\pi kL(H-h)}{\ln(R/r_e)}\left(1+\frac{\ln(r_e/r_i)}{\ln(R/r_e)}\frac{k}{kl}\right)^{-1}$	Ribacchi et al. (2002)	Tunnel lining. Hydrostatic load constant along the tunnel border.
$Q = \frac{2\pi kL(A+D)}{\ln\left(\frac{(H-D)}{r}+\sqrt{\frac{(H-D)^2}{r^2}-1}\right)}$ where $A = (H-D)\frac{(1-\alpha^2)}{(1+\alpha^2)}$ $\alpha = \frac{1}{r}(H-D-\sqrt{(H-D)^2-r^2})$	Park et al. (2008)	Water table above land surface. Hydrostatic load along the tunnel border depending on the stage.
$Q = 2\pi kL\frac{\lambda^2-1}{\lambda^2+1}\frac{(H-h)}{\ln\lambda}$ where $\lambda = \frac{(H-h)}{r}-\sqrt{\frac{(H-h)^2}{r^2}}-1$	El Tani (2003)	Water table below land surface. Hydrostatic load along the tunnel border depending on the stage. Extension for non-horizontal water table.
Transient state		
$Q_t = \frac{4\pi kL(H_t-h)}{\ln(2.25kLt/Sr^2)}$	Jacob and Lohman (1952)	Hydrostatic load constant along the tunnel border
$Q_t = 2\pi\int_0^{vt}\frac{k(H_t-h)\theta(L-x)}{\ln\left[1+\sqrt{\frac{\pi k}{Sr^2}\left(t-\frac{x}{v}\right)}\right]}dx$	Perrochet et al. (2005)	Hydrostatic load constant along the tunnel border Extension for heterogeneous aquifer by Perrochet et al. (2007)

k = hydraulic conductivity; L = length of the tunnel; h = water table into the tunnel; S = coefficient of storage; r = tunnel radius (with lining: r_e is the external radius and r_i in the internal radius); H = initial water table pressure at the tunnel axis; R = radius of influence; k_l =tunnel lining hydraulic conductivity; D = hydraulic load above land surface; s = drawdown above the tunnel (*H-h*); t = time; x = spatial coordinate along the tunnel axis with an origin at the entry of the permeable zone; v= drilling speed;θ (*L-x*) = Heaviside step function (also named unit step function; when $(L-x) < 0$, $\theta(L-x) = 0$; when $(L-x) > 0$, $\theta(L-x) = 1$).

Nevertheless, in order to keep into account also the typical anisotropy of rock masses (Gattinoni et al., 2008), it is better not to refer to the radius of influence, but to the *ellipse of influence*, whose shape can be easily obtained on the base of the hydraulic conductivity tensor (Fig. 6.18).

Actually, knowing the hydraulic conductivity tensors on structurally and geologically homogeneous areas, it was possible to reconstruct the shape of the ellipse of influence on the horizontal plane, with the relevant values and direction of maximum (K_{max}) and minimum (K_{min}) hydraulic conductivity. In the example given in Fig. 6.18a, as K_3 is vertical, the semiaxes of the ellipse in the horizontal plane

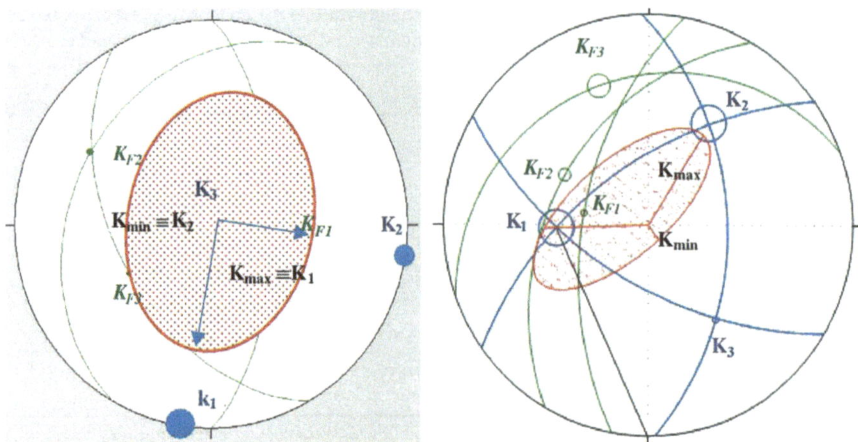

Fig. 6.18 Stereographic projection of the hydraulic conductivity tensor in the rock mass (*in blue*) and of single hydraulic conductivity oriented along the discontinuity planes (*in green*). The dimension of *arrows and colored circles*, representative of hydraulic conductivity vectors, is proportional to the modulus of hydraulic conductivity itself. The shape of the ellipse of influence on the horizontal plane is marked in *red*. (**a**) If a component of the K tensor coincides with the axis *Z*; (**b**) if the components of tensor K have any direction

are oriented along K_1 and K_2, which are, therefore, coincident with the K_{max} and K_{min} assessed on the horizontal plane. In the example of Fig. 6.18b, on the contrary, the representative ellipsoid of the hydraulic conductivity tensor had to be projected on the horizontal plane (the red lines represent the projection of the three semi-axes) and, therefore, the module and the direction of K_{max} e and K_{min} could be determined.

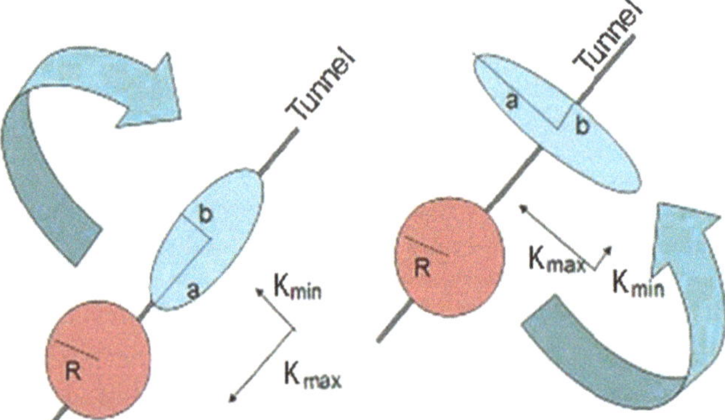

Fig. 6.19 Anisotropy effect in the medium rock on the radius of influence of the tunnel: (**a**) K_{max} is parallel to the tunnel axis; (**b**) K_{max} is orthogonal to the tunnel axis

Starting from the average radius of influence *R*, estimated for the isotropic medium, the semiaxes of the ellipse of influence were obtained (Fig. 6.19) as function of the anisotropy ratio *K*:

$$a = R\sqrt{\frac{K_{\max}}{K_{eq}}} \Rightarrow \quad major\ semiaxes,\ b = R\sqrt{\frac{K_{\min}}{K_{eq}}} \Rightarrow \quad minor\ semiaxes\,.$$

It was observed that the orientation of the main discontinuities determines a marked orientation of the hydraulic flow, bringing about a deformation of the radius of influence. The higher the anisotropy ratio, the higher the deformation. The following relation is valid:

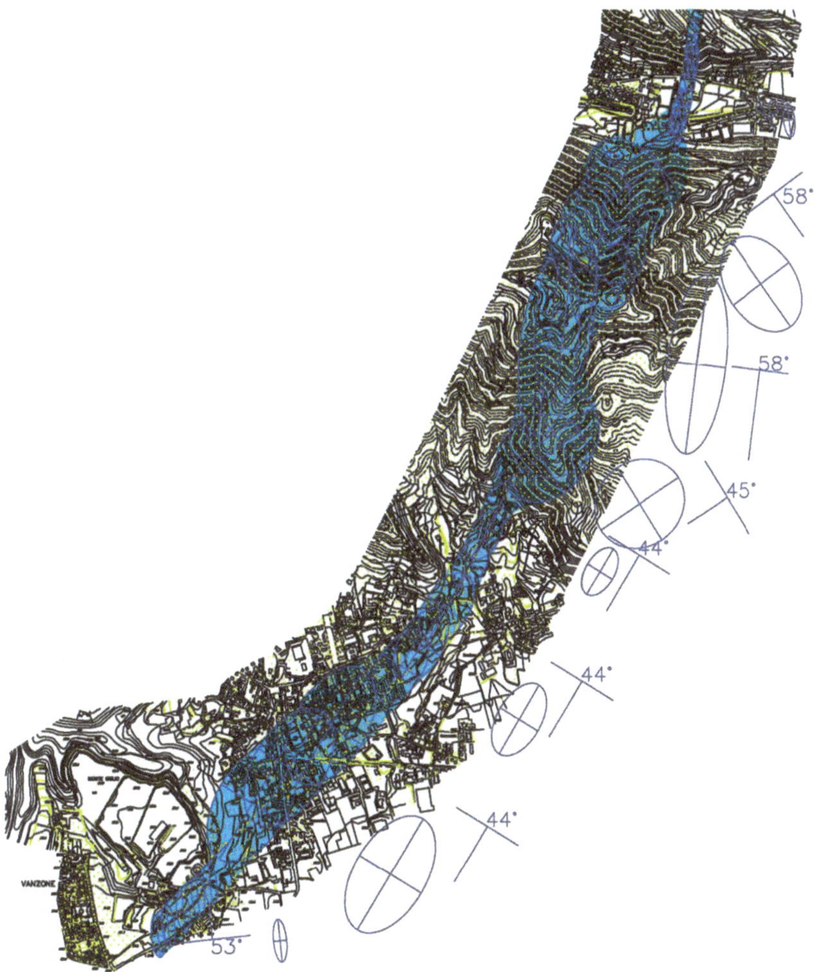

Fig. 6.20 Example of delimitation of the influence zone of a tunnel on the deep aquifer (*in light blue*), according to the rock mass anisotropy (*highlighted by the navy blue ellipses*)

$$\frac{a}{b} = \sqrt{\frac{K_{\max}}{K_{\min}}}.$$

In this way it was possible to bound the area of the aquifer that was potentially influenced by the draining of the tunnel along its whole length (Fig. 6.20). As can be observed, the influence zone has a variable width along the whole plan and it is not symmetrical to the tunnel axis, due to a high dip (generally over 45°) of the discontinuities in the rock mass. In order to check the previously obtained results, it is also possible to use a mathematical model that can simulate the entity and the direction of tunnel inflows. To that aim, a tridimensional flow model to the tunnel scale can be implemented (scale of hydrogeological basin), which allows the simulation of the groundwater flow before and after the tunnel construction, thus evaluating its draining effect (Fig. 6.21). This kind of approach makes it possible to bound the area surrounding the tunnel that is potentially interested by piezometric drawdown when supply conditions change (e.g. rainfalls) and due to different modifications of the original project (e.g. tunnel lining and depth).

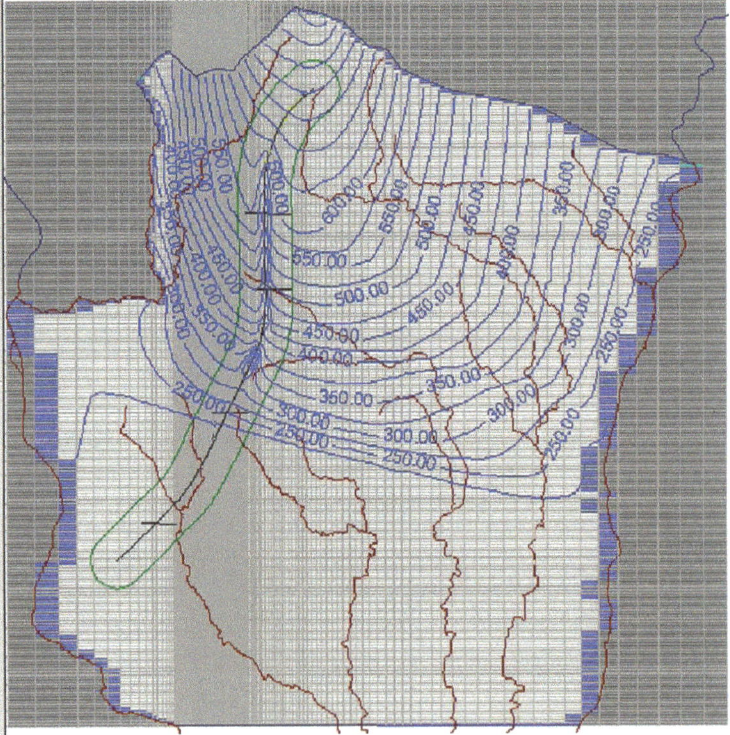

Fig. 6.21 Piezometry of the deep aquifer simulated with the large-scale model. The boundary conditions of constant load and the isopiezometric lines are *blue* navy, the rivers are brown. The *green line* circumscribes an area of 500 m around the tunnel

The marked dependence of the influence zone from the discontinuity network can be evaluated using more detailed numerical modeling, models with distinct elements that can simulate the water flow within the discontinuities. This approach allowed the reconstruction of the flow network induced by the opening of the tunnel and the determination of the flows, as well as the water pressures in each single discontinuity when their orientation changes. In particular, a lack of symmetry was observed in the flow conditions (Fig. 6.22a and b), with values of the influence radius which can also double if there are discontinuity systems with such an orientation (if the other features are equal) that it can favor the water inflow in the tunnel.

The delimitation of the tunnel influence zone thus obtained refers to the deep aquifer within the rock mass that is directly affected by the tunnel excavation. As far as the superficial aquifer is concerned, located inside the quaternary deposits, it is very difficult to define an influence zone that is continuous along the whole tunnel

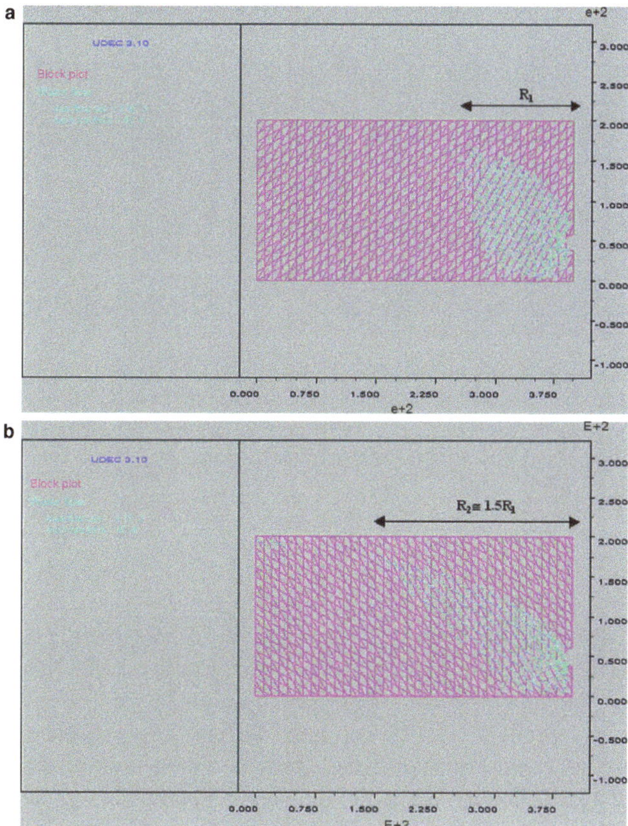

Fig. 6.22 Trend of the inflows within the discontinuities after a tunnel excavation and the relevant radii of influence, when the discontinuity orientation changes. Note that (**a**) and (**b**) represent the two symmetric situations with respect to the tunnel axis

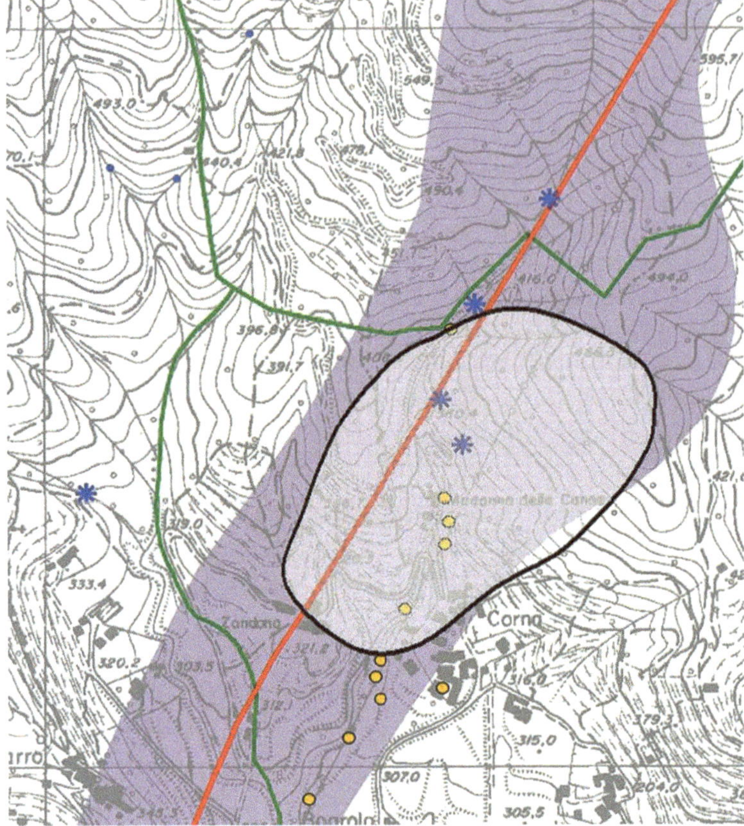

Fig. 6.23 An example of delimitation of the influence zone of a tunnel on the deep aquifer (the *darker color*) and on the superficial aquifer (the *lighter color*, with black contour). The tunnel layout is marked in *red*, the *yellow circles* show the position of piezometers and the *blue asterisks* the springs

axis. That is due to the differences in the quaternary deposits and in their thickness, as well as the different draining relationship with the deep aquifer. In order to bound the influence zone triggered by the tunnel on the aquifer, the interconnection areas between the two aquifers must be delimited and, as a consequence, the draining effect of these interconnection zones on superficial waters must be quantified using, for example, a modeling approach. In this way, it is possible to recreate the main mechanisms ruling locally the drained discharge from the deep aquifer with and without the tunnel and to determine the extension of the catchment area according to the hydraulic features of the interconnection zones, when the boundary conditions change (Fig. 6.23).

6.2.4 Hydrogeological Risk Analysis

The aim of a risk analysis is to assess, in statistical terms, if and to which extent the tunnel excavation constitutes a risk for the hydrogeological setting of the area, in terms of drawdown of the intercepted aquifers or of depletion of the springs, calculating the probability that the tunnel inflow or the piezometric drawdown may exceed a limit value (Francani et al., 2005; Gattinoni and Scesi, 2006). To this aim, the probability distributions of the hydraulic conductivity (one for each homogeneous zone) were reconstructed and Monte Carlo simulations were used to define the variability field of the flows drained by the tunnel and to estimate the probability distribution of the lowering as time goes by. In the latter case, a *performance function* was written as the hydraulic balance on a control volume that is bounded by the land surface (on the top), by the tunnel (on the bottom) and by the radii of influence of the tunnel (sideways) and is expressed as

$$\frac{\Delta V_S}{\Delta t} = i - q_{S-D},$$

$$\frac{\Delta V_D}{\Delta t} = q_1 - q_2 - q_{tun} + q_{S-D},$$

where

- V_S and V_D = the water volumes of the superficial and deep aquifer, respectively;
- i = infiltration rate, which can be estimated starting from the statistical distribution of rainfalls (Fig. 6.24);
- q_1 and q_2 = flows of the deep groundwater in the section above and beneath the control volume, respectively, which can be obtained from the results of the numerical modeling or from the hydrogeological balance;
- q_{S-D} = flow exchanged between the two aquifers, whereas q_{tun} is the tunnel inflow; both are calculated through analytical formulas that were previously illustrated according to the statistical distribution of some terms, with specific reference to the rock mass hydraulic conductivity.

Hydraulic conductivity values were estimated through in situ tests that were integrated with the use of relations based on data obtained from geological–structural surveys (both superficial and in boreholes), such as aperture and discontinuity frequency (Snow relation, Kiraly, Louis relations, etc.). In this way, it was possible to obtain a sufficiently large set of hydraulic conductivity data (Fig. 6.25) that can be used for the statistical tests; the tests provide good results only if they are carried out on data sets that are homogeneous in terms of lithology and structural setting.

The tunnel inflow is a function of both hydraulic conductivity and the piezometric drawdown triggered by the excavation. Theoretically, it would be suitable to consider the drawdown as an uncertain variable and, if needed, to evaluate its

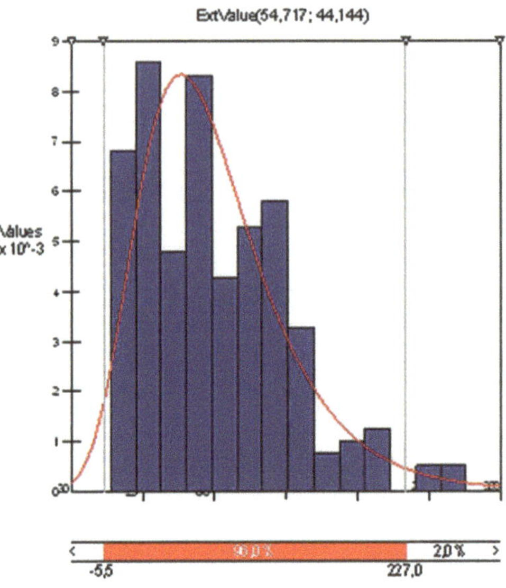

Fig. 6.24 Gumbell's is the most suitable probability distribution to describe the term "infiltration" (Extreme Values)

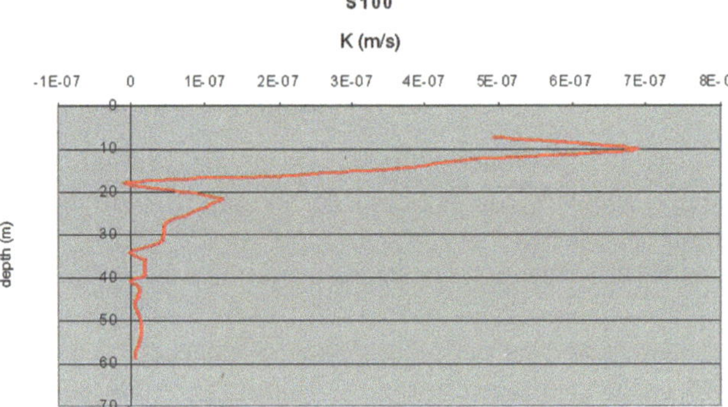

Fig. 6.25 Example of hydraulic conductivity change in depth in a borehole

correlation with the rock hydraulic conductivity. In reality, the available data are often insufficient for a study of this kind and therefore it is necessary to consider the drawdown as a constant that is estimated precautionarily and for homogeneous sections. Considering that hypothesis and starting from analytical formulas, it was possible to reconstruct the probability distribution of the tunnel inflow, thus obtaining the average expected flow and the probability that a certain flow may occur (Fig. 6.26).

The assessment of the hydrogeological risk connected to tunnel construction has to be carried out considering two different points of views

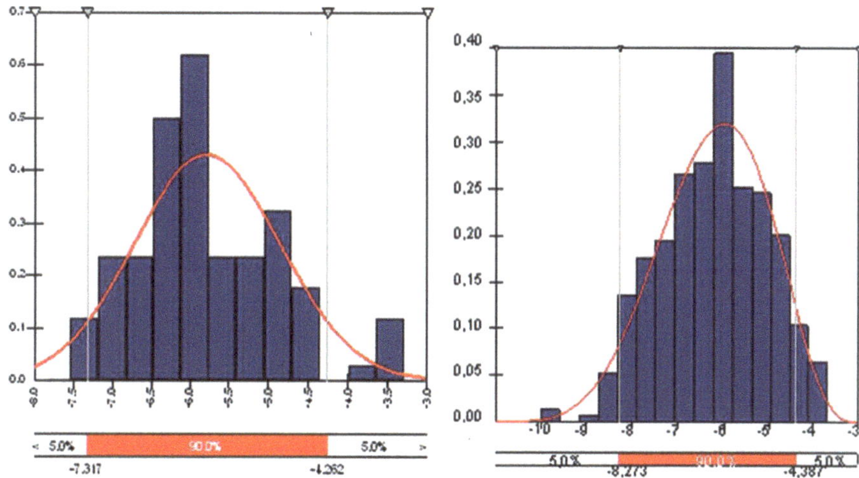

Fig. 6.26 Example of probability distribution of the hydraulic conductivity logarithms in two homogeneous zones: **(a)** in good quality rock mass, the beta generalized distribution is the one that better approximates the sample (p-value 70% and 50% < a < 75%); **(b)** in a shear zone, the normal distribution was adopted

- that of the work (therefore of those responsible for the project or the building firm), in which case it is important to assess the probability of tunnel inflows exceeding the water discharge capacity of the tunnel drainage system (R_{H2O});

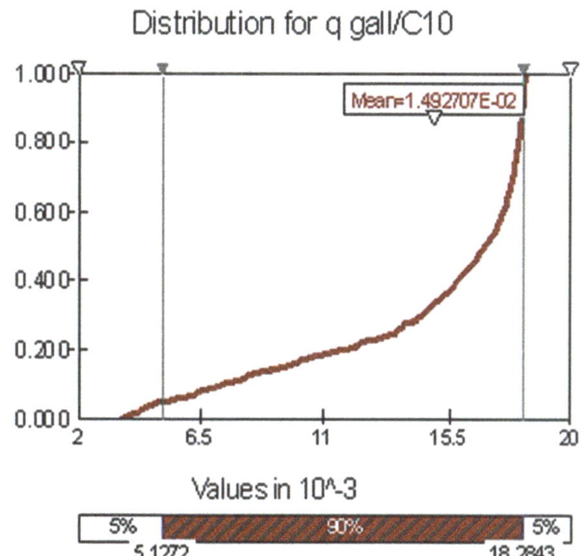

Fig. 6.27 Example of probability distribution of tunnel inflows (expressed in cm/s) in a homogeneous section with average overburden (about 100 m)

- that of the environment (therefore of the population or of public administrators); in this case it is extremely important to foresee the change in the water availability in the area, for example, in terms of drawdown of the water table ($R_{\Delta h}$).

From the point of view of the tunnel, the risk analysis refers to water inflows in the tunnel that could compromise its functionality or safety. In this case, the analysis of probability distribution obtained for the tunnel inflow (Fig. 6.27) can provide useful information to those responsible for the project that would support them in choosing the most adequate measures to limit the risk:

$$R_{H_2O} = p\left[Q\left(t\right) \geq Q_{\text{lim}}\right],$$

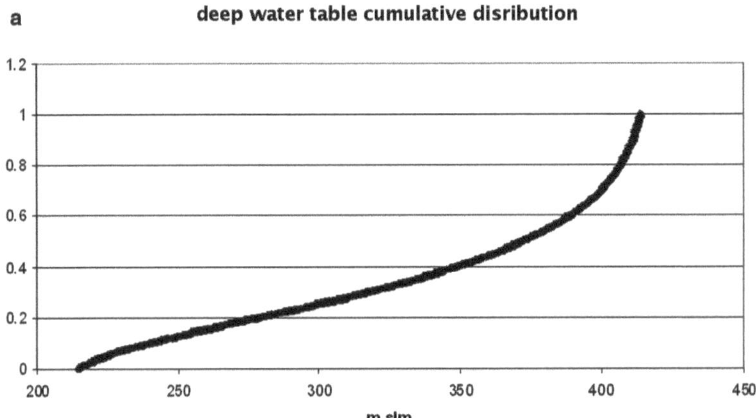

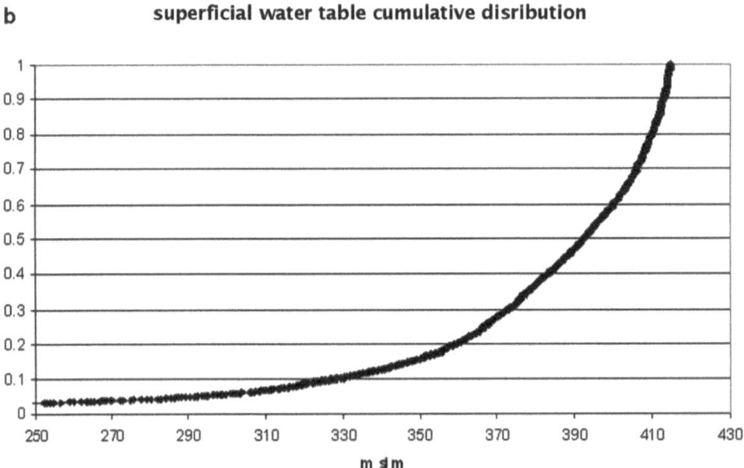

Fig. 6.28 Probability distribution (**a**) of the deep water table when the steady state has been reached; (**b**) of the superficial water table, when the steady state has been reached

as $Q_{\lim}$ is a parameter in the tunnel project that is strictly linked to the dimensioning of the draining system.

Considering the risk analysis from the point of view of the environment, an acceptable drawdown value, Δh, has to be set, which would then allow to define the risk in function of the geological, morphological, climatic features and of the use of the water resources $(x_1, x_2, \ldots, x_n)$ in the area examined as

$$R_{\Delta h} = p\left[\Delta h = g\left(x_1, x_2, \ldots, x_n\right) < \Delta h_{\lim}\right].$$

To this aim, it was possible to carry out the simulation of different risk scenarios starting from the performance function, considering the probability distribution of the piezometric drawdowns induced by the tunnel in the different aquifers (Fig. 6.28a, b). To establish an acceptable drawdown value, in terms of availability of superficial water resources, the average flow of springs was considered; an acceptable piezometric drawdown brings about a 20% flow reduction. In the case in the example, a 10 m drawdown is considered acceptable; the probability to exceed it is equal to 83%. The springs might dry up if a piezometric drawdown equal to 25 m would take place; the risk that this might happen is equal to 18%.

6.3 Hydrogeological Risk Linked with Road Construction

The third example concerns the hydrogeological studies carried out along the Provincial Road that links S. Giacomo to the main road S.S.39 near the Aprica Pass (in the Sondrio province, Northern Italy), which is interested by many instability phenomena. The fall and slide of materials from the slopes, that sometimes determine very high risks for the people traveling along that road, are partly due to the geological, geomorphologic and tectonic setting of the area, and partly to the water flow in rock masses outcropping along that road.

From the geological point of view, the slopes facing the road are made by lithotypes belonging to the Edolo Schists formation, which is constituted by silver-grey micaschists with large garnet crystals. Often the Edolo Schists present inserts of amphibolite and quartzite. The latter, near Aprica, takes the shape of huge lithologically homogeneous masses (Fig. 6.29).

Geological–structural and geomechanical surveys carried out along the whole road axis led to the identification of

- schistosity planes that, together with amphibolite lens inserts, mainly have E–W strike, North dip direction and dip ranging from high to subvertical;
- three discontinuity sets: H_1 (strike NE–SW, dip direction toward south and dip ranging from low to medium); H_2 (strike NW–SE, dip direction toward south or north and dip ranging from medium to high); H_3 (strike mainly N–S, dip direction toward east or west and high dip) (Fig. 6.30);
- the main features and geomechanical parameters concerning both the rock matrix and the discontinuities.

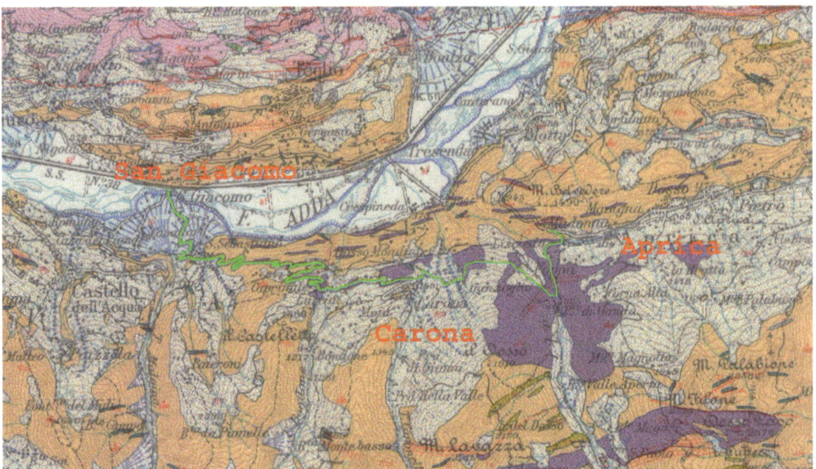

Fig. 6.29 Geological map of the road stretch under examination. In *yellow* are the Edolo schists, in *violet* the amphibolite

The detailed examination of the slopes along the road allowed the identification of areas interested by possible sliding phenomena (toppling, planar or wedge sliding) (Figs. 6.31, 6.32 and 6.33) and of the rock masses characterized by low-quality indexes. The geomechanical classifications by Bieniawski and Romana (Table 6.6 , Romana, 1985) were used to determine these indexes.

As observed in Fig. 6.33, there are sections of the slope that are interested by planar slidings, others by wedge sliding, still others by toppling, and there are

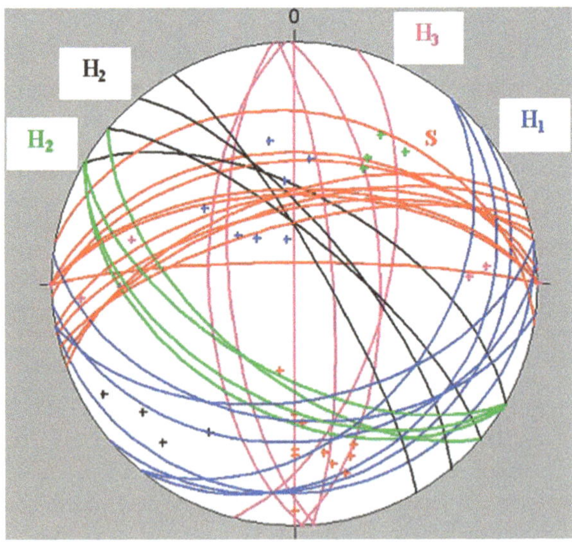

Fig. 6.30 Stereographic representation of the main discontinuity families

Fig. 6.31 Discontinuity families that divide the rock in blocks and possible direction of the movement

some outcroppings (n. 4, 6 and 11) whose quality ranges from low to very low. Table 6.6 clearly shows how, according to Bieniawski classification, the rock quality is generally good (RMR > 61 except in correspondence to the outcropping n. 6), whereas Romana correction allows to differentiate the outcroppings according to their tendency for instability.

The examination of rain gage data highlighted how the area is subject to intense rainfalls that, mainly in spring and fall, may cause localized water flows invading the road and, in colder months, the formation of dangerous sheets of ice.

Moreover, water circulation within rock masses may favor the triggering of slides. For that reason, geological–structural data are used to calculate the hydraulic conductivity values referred to the more superficial layers of rock masses and the main flow directions, considering the interconnection degree among joints.

Being a non-saturated medium, the following formulas were adopted:

$$K = \sum_{j=1}^{m} \left[\frac{e_j^3 \cdot f_i \cdot g}{12 \upsilon} \right] \quad (smooth\ walls)$$

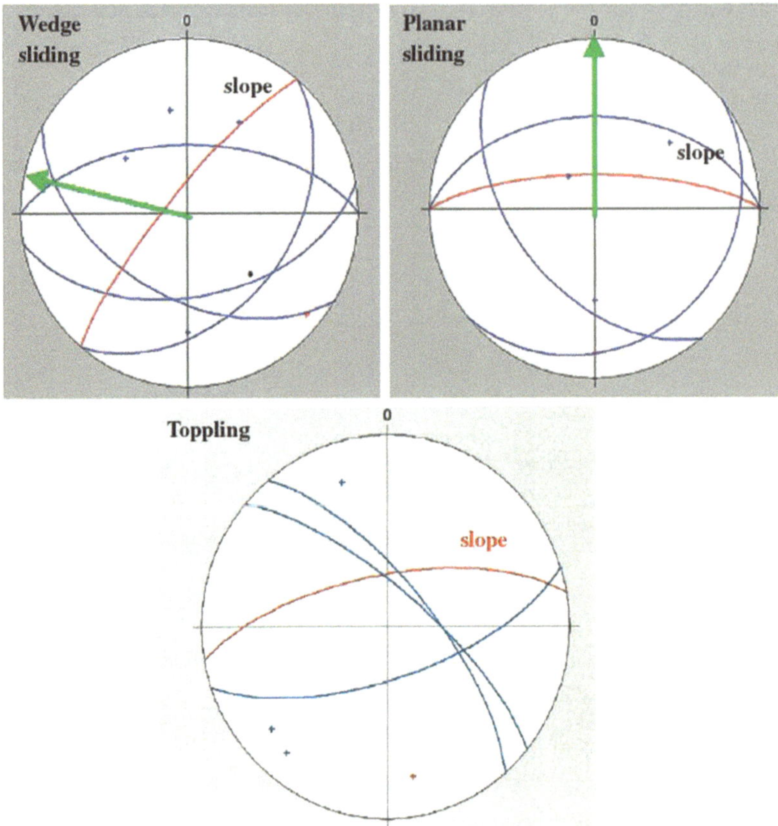

Fig. 6.32 Identification of possible directions of the movement in some outcroppings situated along the road (discontinuities are in *blue*, schistosity as well, the slope is marked in *red*)

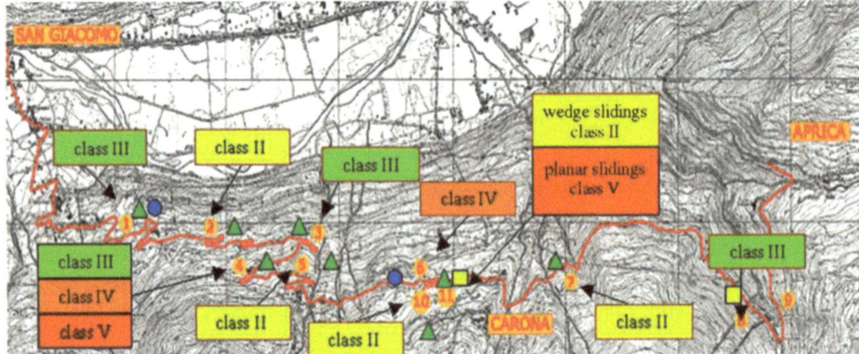

Fig. 6.33 Areas interested by possible wedge slidings (*green triangle*), planar slidings (*yellow square*) and topplings (*blue circle*). Quality classes for rock according to Romana. Outcropping n. 9 is not interested by possible movements and can be placed in Bieniawski III class

Table 6.6 Rock quality index according to Romana and Bieniawski

Outcropping n.	SMR referring to sliding phenomena	SMR referring to toppling phenomena	RMR
1	57.4 (III)	55 (III)	61 (II)
2	65 (II)		74 (II)
3	50 (III)		74 (II)
4	21 (IV)		63 (II)
4	3 (V)		63 (II)
4	12 (V)		63 (II)
4	54 (III)		63 (II)
5	67.35 (II)		75 (II)
5	67.35 (II)		75 (II)
5	67.5 (II)		75 (II)
6		32.75 (IV)	54 (III)
7	72.4 (II)		76 (II)
8	54 (III)		78 (II)
9			66 (II)
10	73.5 (II)		81 (I)
11	61.4 (II)		65 (II)
11	63.65 (II)		65 (II)
11	14 (V)		65 (II)

$$K = \frac{g \cdot e^2}{12\upsilon\left(1 + 8.8\left(0.5 - \frac{e}{2 \cdot JRC^{2.5}}\right)^{1.5}\right)} \quad (\textit{rough walls}).$$

Therefore, for each outcropping, the following information was obtained (Table 6.7; Figs. 6.34 and 6.35):

1. orientation of the discontinuity families;
2. hydraulic conductivity of the different discontinuity families;
3. hydraulic conductivity of the rock mass;
4. main flow direction.

The study about water flow highlighted how, along the whole middle section of the road, the main flow directions converge towards the roadway and how the hydraulic conductivity values of the rock masses are, on the average, quite high.

Table 6.7 Examples of hydraulic conductivity values obtained by structural data

Orientation of discontinuity families	
H_1	N/45°
H_2	140°/20°
H_3	230°/10°

Hydraulic conductivity of the discontinuity families	
K_1 (m/s)	6.54×10^{-3}
K_2 (m/s)	4.18×10^{-4}
K_3 (m/s)	5.23×10^{-5}

Hydraulic conductivity of the rock mass and main direction of the flow	
K_{tot} (m/s)	6.42×10^{-3}
Dip direction (°)	3
Dip (°)	48

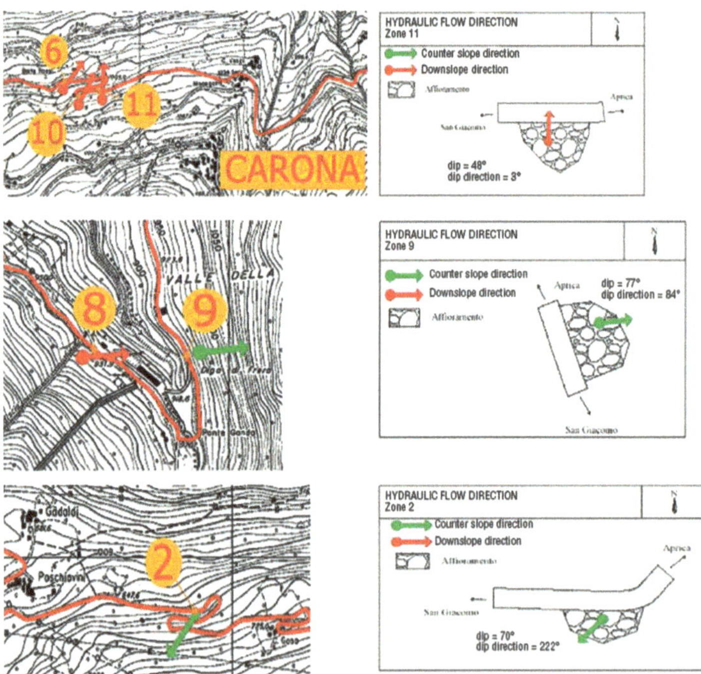

Fig. 6.34 Main direction of hydraulic flow having downslope (*red arrows*) and counter slope direction (*green arrow*) in some outcroppings situated along the road

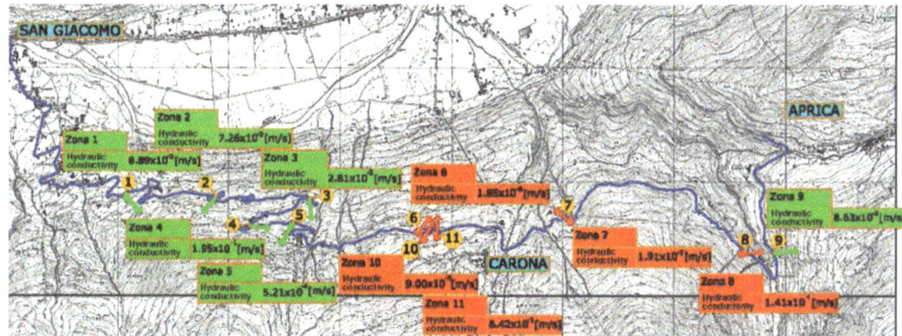

Fig. 6.35 Main flow directions having downslope (*in red*) and counter slope direction (*in green*) and hydraulic conductivity of rock masses at the different outcroppings considered

Table 6.8 Indexes and coefficients used to define the SRM_w quality index

			Planar and wedge sliding		
$(\alpha_d - \alpha_s)$	< 5°	5°–10°	10°–20°	20°–30°	> 30°
F_1 factor	1	0.85	0.7	0.4	0.15
			Toppling		
$(\alpha_d - \alpha_s)-180°$	<5°	°–10°	0°–20°	0°–30°	
F_1 factor	1	0.85	0.7	0.4	–
			Planar and wedge sliding		
i_d	> 45°	45°–35°	35°–30°	30°–20°	< 20°
F_2 factor	1	0.85	0.7	0.4	0.15
			Toppling		
i_d	> 45°	45°–35°	35°–30°	30°–20°	< 20°
F_2 factor	1	1	1	1	1
			Planar and wedge sliding		
$i_d - i_s$	< –10°	–10°–0°	0°	0°–10°	>10°
F_3 factor	60	50	25	6	0
			Toppling		
$i_d + i_s$			> 110°	120°–110°	> 120°
F_3 factor			0	6	25
	Natural Slope	Presplitting	Smooth Blasting	Blasting or Mechanical	Difficult Blasting
F_4 factor	+15	+10	+ 8	0	–8
$(\alpha_{dw} - \alpha_s)$	< 5°	5°–10°	10°–20°	20°–30°	> 30°
F_1w factor	1	0.85	0.7	0.4	0.15
k (m/s)	> 10^{-3}	10^{-4}–10^{-3}	10^{-4}–10^{-5}	10^{-5}–10^{-6}	<10^{-6}
F_2w factor	1	0.85	0.7	0.4	0.15
$i_{dw} - i_s$	< –10°	–10°–0°	0°	0°–10°	>10°
F_3w factor	20	10	5	0	0

Table 6.9 Rock quality indexes according to Romana and Romana-modified classification to take into account also the influence of water

Outcropping n.	SMR referring to sliding phenomena	SMR referring to toppling phenomena	SMR_{slid} + W	SMR_{top} + W
1	57.4 (III)	55 (III)	54.4 (III)	52 (III)
2	65 (II)		63.5 (II)	
3	50 (III)		47 (III)	
4	21 (IV)		19.5 (V)	
4	3 (V)		1.5 (V)	
4	12 (V)		10.5 (V)	
4	54 (III)		52.5 (III)	
5	67.35 (II)		64.8 (II)	
5	67.35 (II)		64.8 (II)	
5	67.5 (II)		64.95 (II)	
6		32.75 (IV)		32.75 (IV)
7	72.4 (II)		72.4 (II)	
8	54 (III)		54 (III)	
9				
10	73.5 (II)		53.5 (III)	
11	61.4 (II)		41.4 (III)	
11	63.65 (II)		43.65 (III)	
11	14 (V)		–6 (V)	

In order to quantify the influence of hydraulic conductivity and the flow directions on the stability of the slopes facing the road, starting from Romana classification another classification was created that would also take into account the hydrogeological aspect. A *quality index* was then identified, called SMR_W, obtained from the following formula:

$$SMR_w = RMR - (F_1F_2F_3) + F_4 - (F_{1w}F_{2w}F_{3w})$$

where (Table 6.8)

- RMR = Bieniawski index;
- F_1, F_2, F_3= indexes of Romana classification that consider the parallelism among the dip direction of the joints and of the slope, the joint dip and the joint dip with respect to the slope dip, respectively;
- F_4= index of Romana classification that considers the slope excavation methods;
- F_1w = index that considers the parallelism between the dip direction of the main flow direction and the dip direction of the slope. It is independent from the type of kinematics, and coefficient similar to those used by Romana for F_1 was applied to it;
- F_2w = index that considers the hydraulic conductivity values referred to as single outcroppings. Also for that index, the coefficient proposed by Romana for F_2 was used;
- F_3w = index that considers the existing relations between the dip of the main flow direction and the slope dip; also this one depends on the type of kinematics. In this case, coefficient capable of reducing the quality index by one class was chosen, to better highlight the effect of water on the stability.

In the case in the example, if the influence of water is also considered, it can be observed (Table 6.9) that the quality of rock masses decreases in some sectors shifting, for example, from the second to the third class or from the fourth to the fifth class. Therefore, the zones requiring focused measures are identified; the measures may be strictly linked to the slope stability or to draining works. In particular, along the road it can be observed that the outcroppings 4, 10 and 11 show a very poor quality and that water might contribute towards aggravating their susceptibility to landslide.

6.4 Mountain Aquifer Exploitation and Safeguard: Eva Verda Basin Case Study (Saint Marcel, Aosta Valley, Italy)

In the last few years, a progressive impoverishment of water resources took place in the mountains due to the increase in anthropic activity and the change in the precipitation regime. In this context, the preservation of every aquifer is extremely important. In particular, mountain fissured aquifers could play a relevant role in the water supply, even though they have a limited productivity. These aquifers represent the main feeding source for deep groundwater in plain areas. Moreover, the mountain aquifers are vulnerable to external factors such as the climate change and pollution. Actually, the springs are located close to their recharge areas, and as a consequence, their circuits are short and swift. A good knowledge of the mountain hydrogeological circuits allows both their safeguard and a better exploitation (Wolkersdorfer and Bowell, 2005).

This example concerns the determination of the groundwater system in an area located on the right side of the Saint Marcel Valley (in the dextral side of the Aosta Valley, Northern Italy). This area includes the Fe–Cu disused mines of Servette and Chuc. The sulfide mineralizations are enclosed within dominant basic rocks (meta-ophiolites) of the Piedmont nappe, characterized by a strong ductile-brittle deformation due to Alpine faults. The highly fractured rocks of the Servette mine represent the main recharge area for the springs located at the bottom of the study area, in the lower Saint Marcel valley (Fig. 6.36).

Two streams flow in the study area: the Saint Marcel torrent, which runs in the N–S direction at the valley bottom, and the Eva Verda stream, which runs in the E–W direction in the centre of the study area. There are also 11 springs: one of these is located in the Servette mine, at the "Ribasso gallery" 1815 m a.s.l.; other springs arise in the bottom of the study area at about 1400 m a.s.l.; only 6 springs are perennial. One of these, situated in the proximity of the abandoned factory of the Chuc mine that was exploited until 1957, shows a brilliant green color and is known as "Eva Verda" among the local inhabitants (Noussan, 1972). The study area is covered by quaternary deposits for 85% and by outcropping rocks for the remaining 15% .

The quaternary deposits are constituted by the following:

- Glacial deposit above 1550 m a.s.l. It consists of blocks of different sizes (ranging from m to dam) within a clayey matrix. A NNW–SSE trending moraine cord also occurs between 1450 m and 1410 m of altitude.
- Landslide deposits occur between 2000 m a.s.l. and 1350 m a.s.l. in the Servette slope. The widest paleo-landslide deposit (about 700 m^2 of extension) is located along the Servette–Praborna country road. This deposit is rather heterogeneous and characterized by metric blocks of glaucophanite, talcschists and minor prasinite that are homogenously distributed (coarse landslide deposit). By contrast, minor landslide deposits are constituted by decimetric blocks of prasinite and serpentinite (fine landslide deposit).
- Colluvium forms a few decimeters thick but extended deposit is covered by vegetation.
- Anthropic deposits are present in the northern part of the study area and near the powder magazine and the cableway that joined Servette to Chuc. These consist of massive slags and other material derived from diggings of the exploited galleries. The most extended deposit is about 15 m thick. Slags are centimetric in size and constituted by silica, oxides and sulfides.

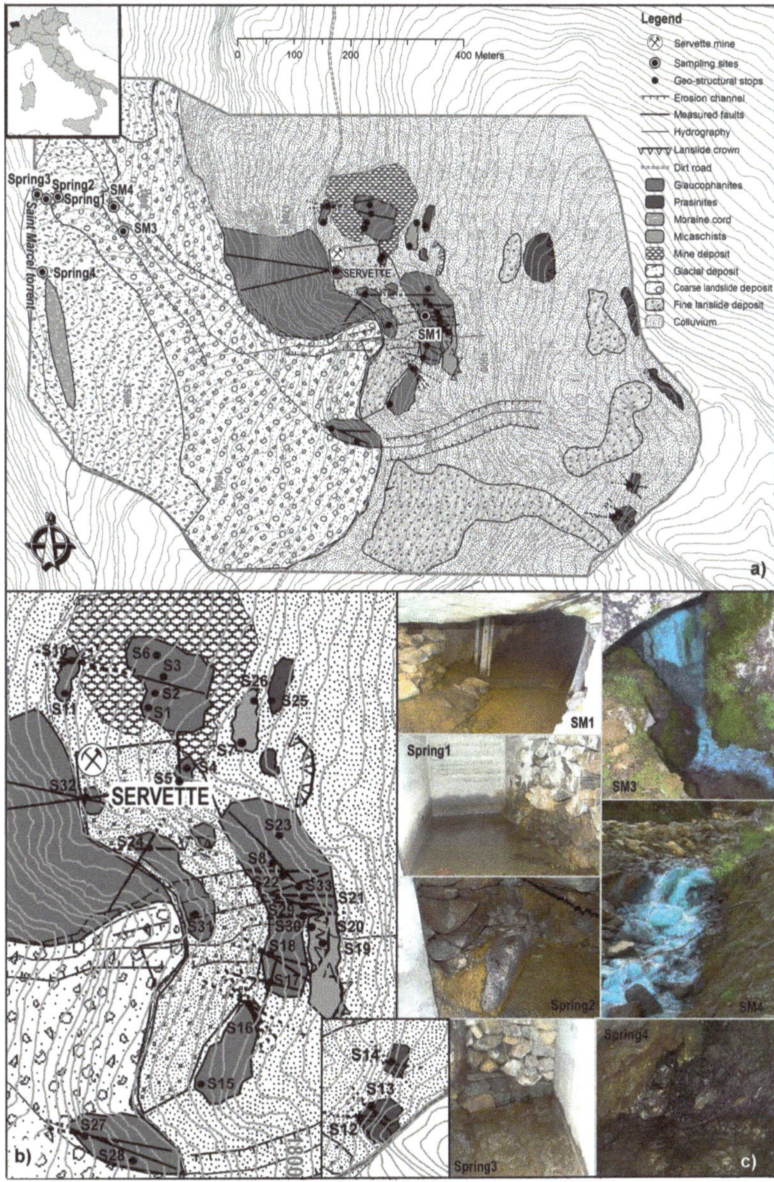

Fig. 6.36 Study area map with geo-structural stops and sampling sites location. (**a**) Geological map with sampling site locations; (**b**) geostructural stop locations; (**c**) pictures of sampling sites

The outcrops are represented by (Fig. 6.36a) the following:

- Glaucophanite with frequent intercalation of talcschists. They are dominant in the central part of the study area between 1500 m and 1820 m a.s.l. The rocks are locally altered by the acid drainage water. Talcschists layers outcrop along the country road from Servette to Praborna (Krutow-Mozgawa, 1988).

- Prasinite, metasomatic rocks (massive fels and rodingite) and serpentinite. They crop out at an altitude of 2000 m (at the very top of the study area). Metasomatic rocks are visible below the serpentinite at 1815 m a.s.l. The outcrops are scattered and discontinuous due to the presence of the colluvial deposit.
- Micaschists. A thin layer crops out at the point of contact with metasomatic rocks, at 1815 m a.s.l., above the uppermost tunnel entrance, and at the point of contact with glaucophanite in proximity of the "Ribasso gallery" entrance (1780 m a.s.l.; southern sector).

The Saint Marcel Valley is oriented N–S along a tectonic lineament. The regional schistosity (290°/40°) is deformed by folds with N–S trending axes. The right slope of the valley, where the Servette mine is located, is characterized by a down-slope orientation, strongly influenced by the tectonic structures. In the Saint Marcel Valley there are two main fault systems: a brittle N–S trending fault (Martin et al., 2007; Martin et al., 1989) and the E–W trending Aosta-Ranzola fault (Tartarotti et al., 1986; Gianotti, 1999), which were responsible for the deformation observed in the Servette mining area.

The Servette slope is characterized by NE–SW trending mylonites, well recognizable along the Servette–Praborna country road, and by several fault planes related to the Aosta-Ranzola fault (E–W direction), and finally by considerable fracturation with local slope tensioning. The latter caused the opening of numerous fractures with width ranging from 1 mm to more than 1 cm, with strike mostly N–S and subparallel to the slope. Fractures show different filling represented by musk and soil in the prasinite, by clay derived from rock weathering and musk in the glaucophanite and micaschists.

For this work, 1500 discontinuities and 40 fault planes were measured and reworked.

6.4.1 Hydrogeological Reconstruction

The hydraulic conductivity calculation was done using the geo-structural data collected at 31 structural stations only in the visible rocky portion of the Servette slope (Fig. 6.36b). These data were used to calculate the main flow directions, the hydraulic conductivity values (K, m/s) in the unsaturated zone (Table 6.10) and the hydraulic conductivity tensor (K_{tens} m/s) in the saturated zone (Table 6.11). According to preliminary seismic survey, along the Servette slope, the boundary between the unsaturated and saturated zones is presumably located at a depth of 100 m in the rocks.

The aperture reduction of fractures with depth was not taken into consideration for the hydraulic conductivity calculation in the saturated zone, because of

Table 6.10 Hydraulic conductivity values around unsaturated zone (m/s) – dip and dip direction of the hydraulic conductivity vector

Stop	Dip direction (°)	Dip (°)	K_{tot} (m/s)
1	233	71	2.31E+00
2	63	53	3.35E-03
3	180	61	1.51E-03
4	175	70	8.56E-01
5	163	59	2.12E-01
6	107	84	2.70E-02
7	38	35	1.35E-02
8	190	65	1.40E-02
10	191	64	1.34E-02
11	200	65	2.36E-01
15	283	60	5.15E-01
16	94	52	20.54E-03
17	330	36	4.66E-01
18	40	34	9.33E-02
19	336	73	2.80E-03
20	200	64	1.36E-03
21	106	83	5.01E-03
22	94	78	4.34E-03
23	206	56	1.84E-03
24	267	83	3.13E-03
25	74	85	3.17E-04
26	259	54	5.89E-04
27	348	47	4.50E-03
28	188	73	5.09E-03
29	262	58	5.13E-02
30	219	79	9.69E-01
31	200	50	3.41E-06
33	3	43	1.32E-02
SE zone			
12	45	86	2.05E + 00
13	235	73	8.65E-02
14	137	83	3.34E + 01

the reduced thickness of the unsaturated one. On the base of the aperture and the different types of flow a different flow equation was used.

Twenty-eight structural stations are located within the Servette mine zone and three in the south-eastern corner. The hydraulic conductivity values of the unsaturated zone in the mine area range between 3.41E-06 m/s (Stop 31) and 2.31E+00 m/s (Stop1, Fig. 6.37). The higher hydraulic conductivity values correspond to the high fractured rock zones.

The hydraulic conductivity distribution in this area is represented in Fig. 6.37: low hydraulic conductivity areas are marked in light brown; these values are linked to the clay material filling of the fracture due to the rock alteration. High hydraulic conductivity areas are indicated in light yellow; these values are due to the rock fracturation derived by the tectonic control.

Table 6.11 Hydraulic conductivity tensor around saturated zone

	K_1			K_2			K_3		
Stop	m/s	Dip Dir. (°)	Dip (°)	m/s	Dip Dir. (°)	Dip (°)	m/s	Dip Dir. (°)	Dip (°)
1	4.07E-02	9	78	1.40E-04	180	3	1.44E-04	107	36
2	2.36E-05	29	5	2.91E-05	0	69	4.88E-05	120	36
3	7.72E-07	0	61	1.55E-05	0	24	1.62E-05	241	29
4	2.76E-02	360	51	1.69E-01	0	32	1.88E-01	243	38
5	1.04E-02	23	78	3.45E-03	0	7	4.18E-03	241	20
6	4.09E-05	276	84	2.57E-04	180	0	2.85E-04	95	50
7	1.90E-06	218	35	1.34E-04	71	50	1.36E-04	140	21
8	8.98E-06	8	63	1.37E-04	0	24	1.46E-04	254	38
10	8.86E-06	9	63	1.30E-06	0	24	1.39E-06	254	38
11	7.10E-02	346	61	2.24E-01	0	13	2.90E-01	196	26
12	4.50E-06	225	86	2.05E-02	46	4	2.05E-02	0	0
13	2.64E-05	54	72	8.62E-04	42	18	8.75E-04	0	5
14	1.64E-02	318	83	3.26E-01	180	2	3.42E-01	152	8
15	4.65E-04	102	59	4.41E-02	180	21	4.45E-02	151	24
16	6.15E-08	274	52	2.53E-05	148	24	2.54E-05	0	36
17	8.35E-06	0	36	4.66E-03	180	35	4.67E-03	0	89
18	8.65E-07	220	34	9.32E-04	50	56	9.33E-04	133	7
19	4.51E-06	156	75	2.59E-05	0	15	3.04E-05	251	5
20	7.14E-11	20	64	1.36E-05	0	0	1.36E-05	200	27
21	5.85E-06	280	83	4.86E-05	176	2	5.19E-05	0	59
22	3.81E-06	270	77	4.16E-05	180	6	4.54E-05	121	21
23	1.66E-06	19	51	1.88E-05	0	30	2.04E-05	258	63
24	4.99E-07	87	83	3.11E-05	83	7	3.16E-05	0	1
25	3.41E-07	235	85	3.03E-06	0	0	3.35E-06	237	9
26	8.93E-07	79	53	5.47E-06	83	37	6.36E-06	0	2
27	5.35E-06	173	46	4.30E-05	180	10	4.83E-05	153	45
28	1.23E-06	8	73	5.05E-05	0	10	5.12E-05	223	19
29	4.89E-05	84	59	4.95E-04	19	14	5.38E-04	0	49
30	7.46E-03	39	84	1.16E-01	39	6	1.24E-01	0	0
31	2.09E-11	20	50	3.41E-08	17	40	3.41E-08	0	5
33	2.46E-05	2	40	1.28E-04	0	43	1.39E-04	255	54

The flow directions in the unsaturated zone are represented by arrows in the hydraulic conductivity map (Fig. 6.37). They converge to the West, towards the Saint Marcel stream. However, some versus anomalies are visible. They are probably controlled by the local tectonic setting (Stop 5, Stop 16, Stop 21, Stop 22, Stop 24, Stop 3, Stop 2 and Stop 25).

In the south-eastern corner, where the outcropping prasinite is interested by E–W and NW–SE fault planes (sketch in Fig. 6.35: Stop12, Stop13, Stop14), the hydraulic conductivity values are higher than the ones in the mining area (Table 6.10). The flow direction in this area is towards the East and brings out the water from the Eva Verda basin to nearby basins.

Most of the flow directions follow the Eva Verda basin, as indicated in the geological map (Fig. 6.38). Regarding the saturated zone, the K_1, K_2 and K_3 tensor

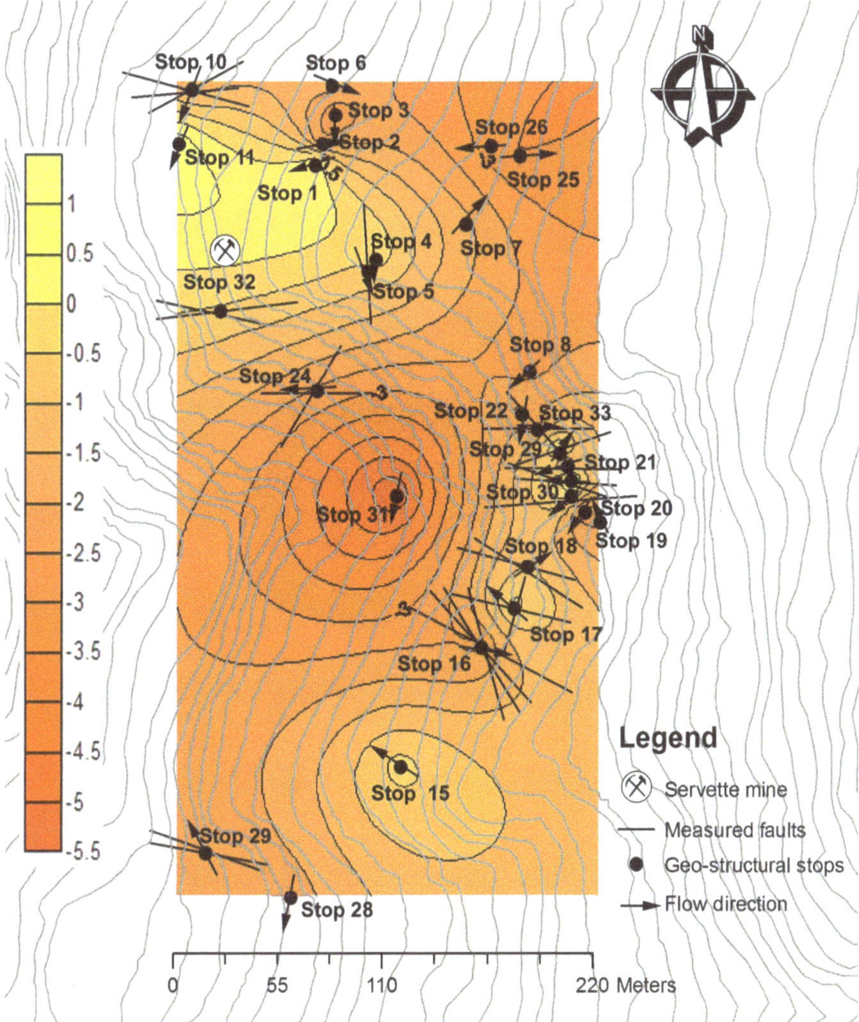

Fig. 6.37 Servette zone hydraulic conductivity contouring: Log *K*

values were calculated (Table 6.11); they show the highest hydraulic conductivity values mainly along the K_3 direction. This one has an apposite direction as regards the hydraulic gradient, rebounding on downhill spring flow. The equivalent hydraulic conductivity (K_{eq}) in the saturated zone was calculated and it corresponds to 6.08E-02 m/s.

Besides the structural analysis, also because of outcropping shortage, a chemical study about the surface and subsurface waters (springs and streams) was carried out.

Seven sites were chosen for sampling (SM1, SM3, SM4, Spring1, Spring 2, Spring 3 and Spring 4; Fig. 6.36c). SM1 spring is located at the entrance of the

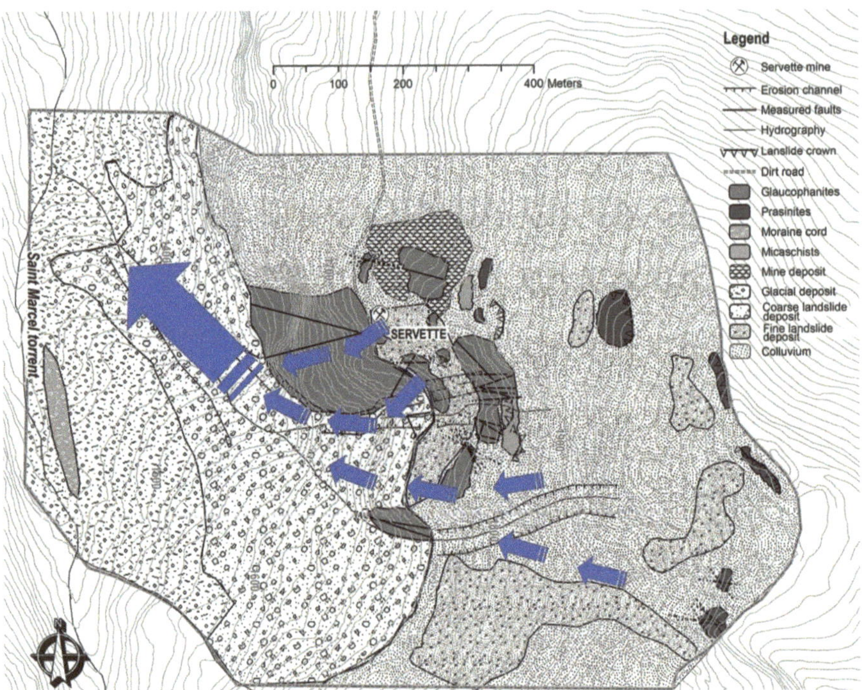

Fig. 6.38 Unsaturated fractured zone flow map (*arrows* represent the main flow direction obtained by structural data)

"Ribasso gallery" and is characterized by the color yellow and mine pollution. The SM3 corresponds to the Eva Verda colored famous spring. The SM4 is located along the stream and has originated from the mixing of the Eva Verda water and an acidic drainage stream. Spring 1, Spring 2 and Spring 3 are perennial springs located downhill SM3 and SM4 sites. At present these springs are intercepted by the Communal Authority. Spring 4 is a fresh water spring located to the South of Spring 1, Spring 2 and Spring 3, in the proximity of the abandoned Chuc mining factory.

The chemical data obtained from water samples highlight the following three groups of water:

- $SO_4{}^{2-}$; Ca^{2+}; Mg^{2+} rich water (SM1);
- $HCO_3{}^{-}$; $SO_4{}^{2-}$; Ca^{2+} rich water (SM3, SM4, Spring 1, Spring 2 and Spring 3);
- $HCO_3{}^{-}$; Ca^{2+} rich water (Spring 4).

The second water group derives from the mixing between the first water that was characterized by the yellow color and was polluted by the mine (SM1) and the third fresh water (Spring 4).

The mixing percentages due to the merging of water from Spring 4 with the second water group are represented in Table 6.12.

The mixing between the waters supplying SM1 and Spring 4 caused a sudden change in the chemical-physical features. The SM1 circuit (blue arrows, Fig 6.39) corresponds to the data obtained by the hydraulic conductivity analysis and the flow direction in the unsaturated fissured zone whose main direction is along the slope (270°) and identifies two water circuits.

The mine drainage water (SM1) leaks in the landslide deposit because it is triggered by high fracturation of the rocks and the persistence of the E–W fault system. The contact between bedrock and landslide deposit represents a very low hydraulic conductivity level. This level blocks the hydrogeological circuit deepening because the fractures are filled up with weathering rock material characterized by fine material and clayey ores.

Table 6.12 Water mixing % by end-member Spring 4

	SM4	SM3	Spring 1	Spring 2	Spring 3
Spring 4 mixing %	64.79	66.90	74.65	77.46	78.87

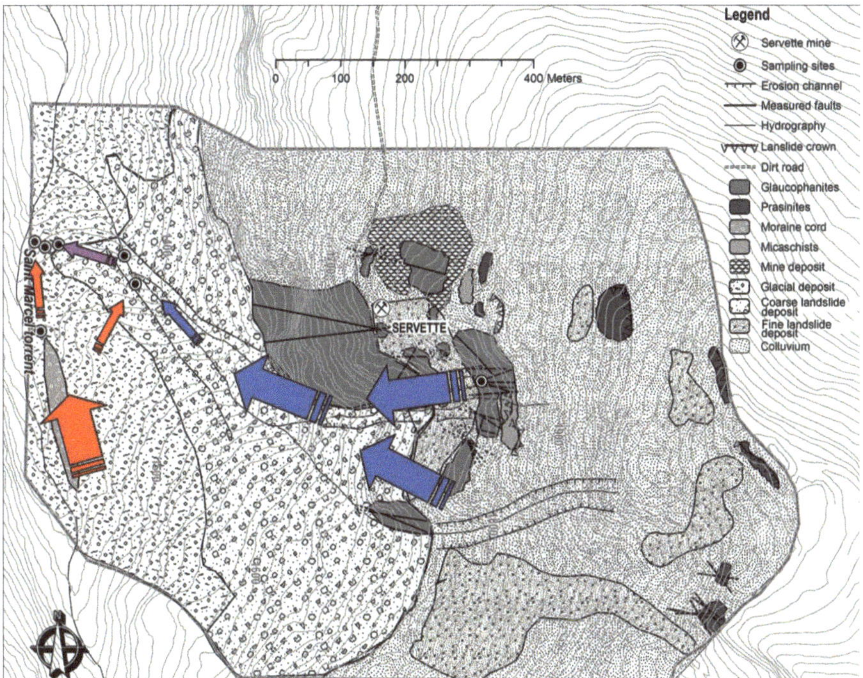

Fig. 6.39 Groundwater system reconstruction. (*red*: SM1 circuit; *blue*: Sorg4 circuit; *violet*: mixing)

In the south-western part of the study area (near Spring 4), the hydraulic conductivity analysis based on geo-structural method could not be carried out due to the lack of outcroppings. Spring 4 circuit, therefore, was determined only by geochemical method. This method points out the water circulation into prasinite (red arrows; Fig. 6.39). The mixing of Spring 4 and SM1 circuits occurs (violet arrows, Fig. 6.39) near SM3 spring ("Eva Verda").

The mine drainage water percentage in Spring 1, Spring 2 and Spring 3 is low (Table 6.12); it derives by the high dilution of Spring 4 water. For this reason, and due to their chemical characteristic (Directive 98/83/EC), the water from Spring 1, Spring 2, Spring 3 and Spring 4 can be used for drink water supply. Besides, the evaluation of the saturated zone hydraulic conductivity pointed out that there is no coincidence between the Servette-Eva Verda hydrographic basin and hydrogeologic basin. The water from the saturated zone flows towards the SE direction taking out water from the Servette-Eva Verda basin and the Saint Marcel basin. This causes the decrease in spring flow as compared to the amount of rainfall.

The integration between geo-structural method used for the hydraulic conductivity determination and the geochemical method allows a full characterization of mountain aquifers and aquifers with complex geological conditions.

6.5 Stochastic Groundwater Modeling for the Drying Risk Assessment

Nowadays, the problem of the supply and protection of the groundwater resources is a highly topical subject. In recent decades, gradual drying up of many springs, low discharge during dry months and perennial springs becoming seasonal were reported all across the Italian Alps and Apennines (Sauro, 1993; Fiorillo et al., 2007; Cambi and Dragoni, 2000). Similar problems related to fractured-karst springs can be found also in many other countries, such as China (Ma et al., 2004; Qian et al., 2006), France (Labat et al., 2002), Germany (Birk et al., 2004), Romania (Oraseanu and Mather, 2000), India (Negi G.C.S. and Joshi V., 2004), Turkey (Elhatip and Gunay, 1998) and the United States (Scanlon et al., 2003).

Such a trend is linked both to the over-abstraction of the groundwater resources and to climate changes. In recent years, many studies were carried out relating the effect of climate changes on surface water (Drogue et Al., 2004; Meenzel and Burger, 2002; Orlando and Byrton, 2003); however, very few researches exist on the potential effects of climate change on groundwater (IPCC, 2001). Available studies show that groundwater recharge and discharge conditions are consequences of the precipitation regime, climatic variables, landscape characteristics and human impact (Allen et al., 2004; Brouyère et al., 2004; Woldeamlak et al., 2007).

Hence, the study of the relationship between rainfall and spring discharge is important to understand the hydrogeological behavior of springs and to optimize the water resources management. The overall aim of the study was to point out a tool able to forecast water resources shortages, as they will probably became more and more frequent in the future. Therefore, the relation between spring depletion curve and the hydrogeological and climatic characteristics of the spring supply field has been studied. A numerical model was selected to study the groundwater flow in the fractured-karsts media in the Nossana Spring field (Bg, Italy), which is one of the most important water resource of the Lombardy Region, for

- simulating the spring regime with different rainfall conditions;
- forecasting its drying risk in terms of probability, considering different climate scenarios.

6.5.1 Hydrogeological Setting of the Study Area

The supply field of the Nossana Spring is located about 70 km Northeast of Milan and covers over 80 km^2. It is a part of the Central Prealps Area. It is characterized by great differences in altitude (the highest altitude reaches 2500 m a.s.l.) and it is made up of carbonatic rocks, for the most part (Chardon, 1974; Bini et al., 2000). The great extension of the supply field explains, at least partly, the high discharge of this spring (on average 3 m^3/s). Furthermore, its regime is characterized by a considerable variability, having discharge swaying ranging from 0.6 to 20 m^3/s. Such a variability results from the karsts nature of the field. Actually, the Nossana Spring is the surface discharge point of a subsurface fractured-karst aquifer (Jadoul et al., 1985) that involves a large fold synclinal in the formation of the Esino's Limestone, very permeable because of karstification, standing in tectonic contact on a marls formation (Fig. 6.40). Groundwater generally flows from NO to SE and the flow is ruled by the topography and faults. Moreover, the low permeability marls formation limits the groundwater flow approaching the South, where the natural upward flow finally discharges to the ground surface and forms the Nossana Spring.

The annual discharge function of the Nossana Spring (Fig. 6.41) is typical of a karsts spring in a mountain area; actually, it shows two peaks:

- a first peak during the spring, due to both the melting of snow and the rainfall;
- a second peak in autumn, resulting from summer rainfall.

During the recharge season, dolines and swallow-holes represent the main infiltration paths for precipitation water, whereas the deeper karsts evidences (i.e. the faults and the fractures) are the main groundwater paths that create a very complex and quite interconnected network. This aquifer supplies the Nossana Spring and it comes to the surface in the last part of the homonymous river (Fig. 6.42). Evidently, during the dry season, the Nossana flow comes from the drainage of the water stored into the karsts pipes.

Due to the Mediterranean climate, which causes a rapid increase in evapotranspiration and a decrease in rainfall in the spring-summer period, the spring discharge amplifies the effect of poor rainfall, especially during drought (Fig. 6.43).

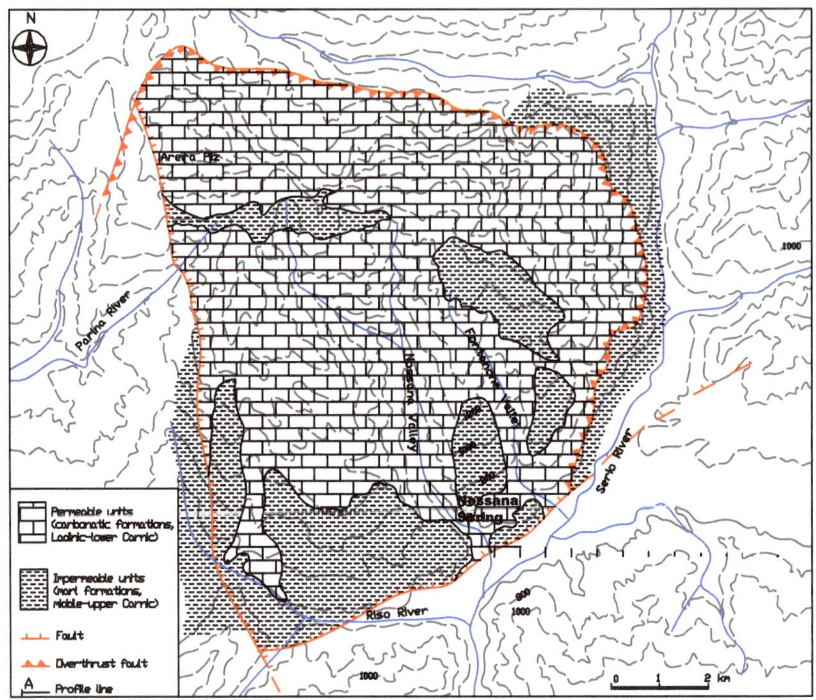

Fig. 6.40 The hydrogeological map of the Nossana Spring catchment area

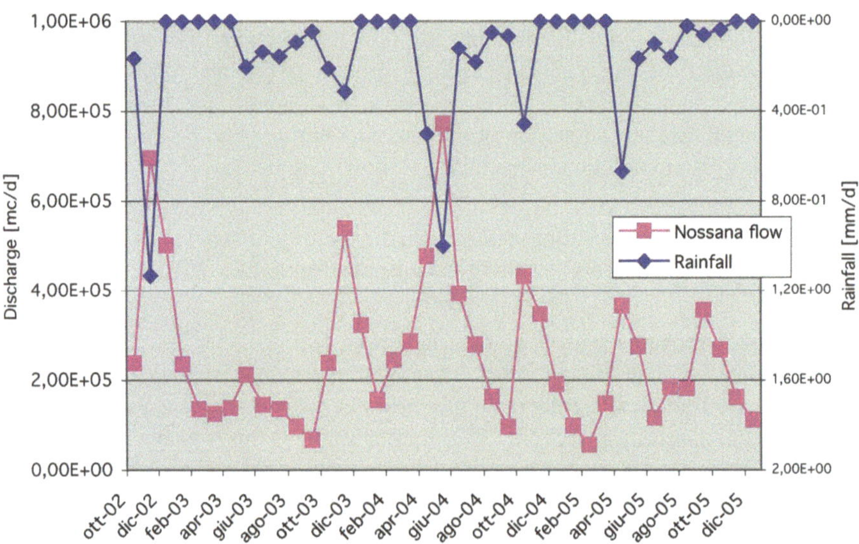

Fig. 6.41 Discharge of the Nossana Spring in the last years, compared with the rainfall

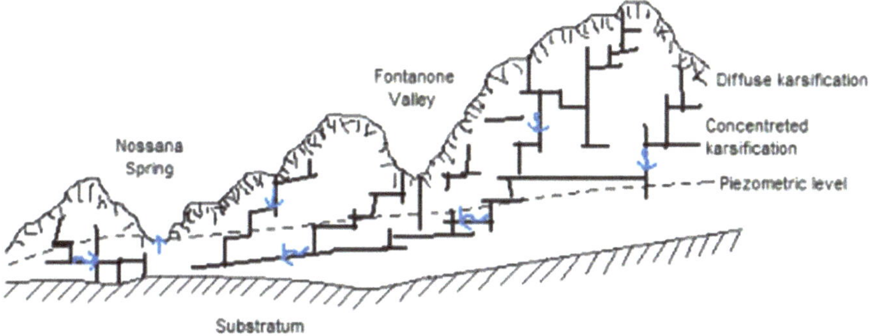

Fig. 6.42 Hydrogeological sketch of water supply to the Nossana Spring

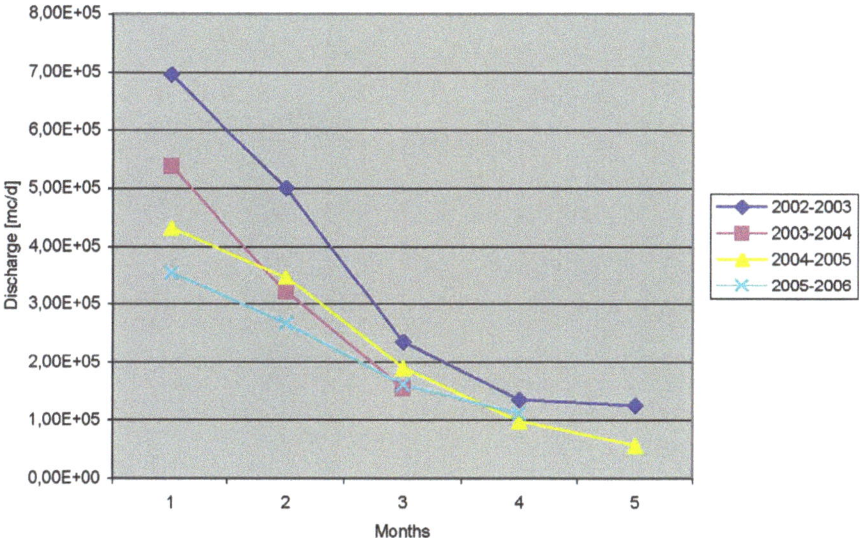

Fig. 6.43 Depletion curves of the Nossana Spring in the last few winters

6.5.2 Groundwater Model of the Nossana Spring

Based on the geological and hydrogeological characteristics of the area, a three-dimension transient-state flow model is presented in the following paragraphs, using the numerical model MODFLOW (Harbaugh et al., 2000).

For a large flow scenario such as this study, treatment of a fractured-karst medium as homogeneous might lead to a reasonably good agreement with an equivalent porous medium for three reasons in particular:

- The conduit systems supplying the spring are fairly uniformly distributed and well-interconnected (Croci et al., 2003);
- the aim of the study is to assess the spring discharge (Scanlon et al., 2003);

- information is available only on spring discharge for model calibration (Angelini and Dragoni, 1997).

The Nossana field was split in 240 ∞ 200 square cells having the maximum size of 50 m (Fig. 6.44). Some topographic variations can be seen from the cross-section diagram of Fig. 6.44. To simulate decreasing hydraulic conductivity and porosity in depth, 7 layers having different thickness were considered.

As no data are available on the hydrogeological parameters, the hydraulic conductivity map (Fig. 6.45) was drawn on the basis of the hydrogeological setting and karsts evidences.

As far as boundary conditions are concerned, the following conditions were applied (Fig. 6.46):

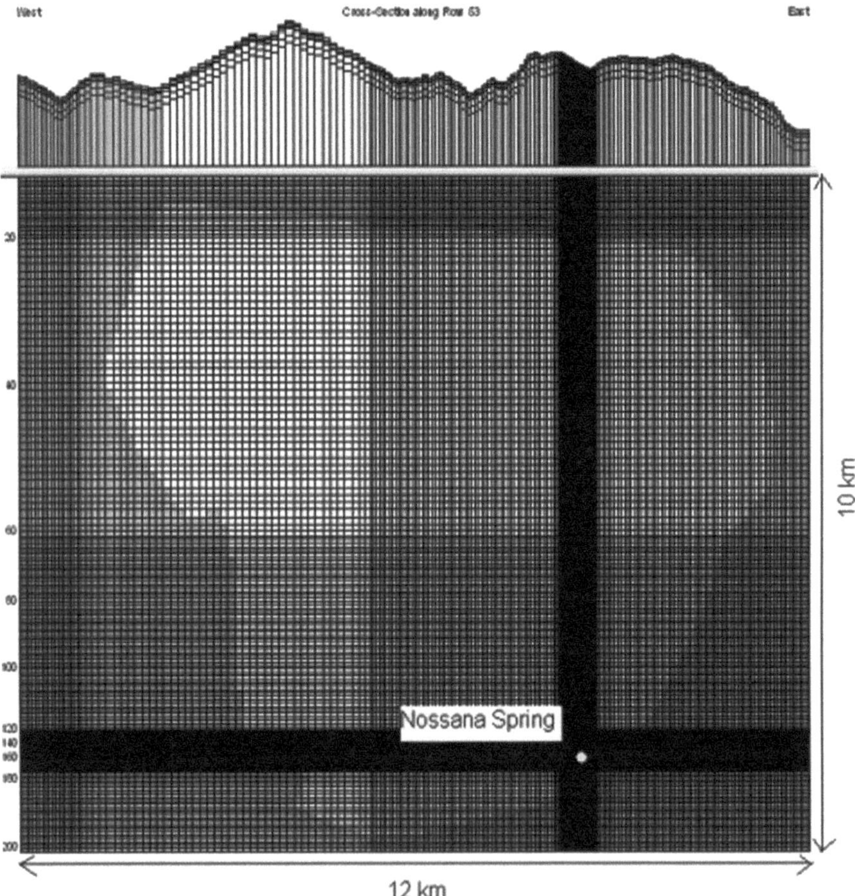

Fig. 6.44 The mesh of discretization: above an E-O cross-section and below the plan view

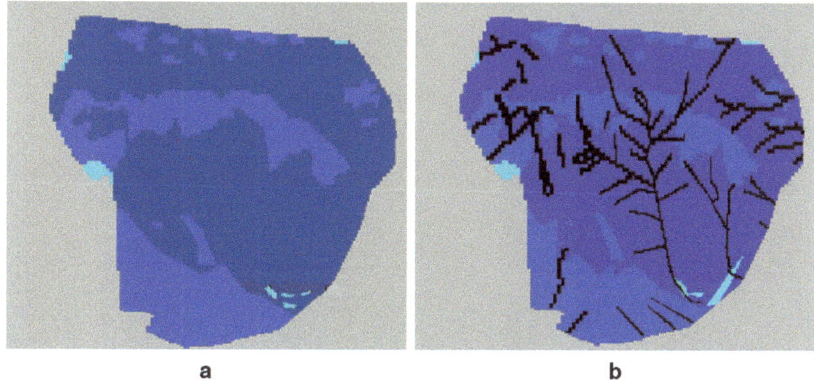

Fig. 6.45 Hydraulic conductivity map: (**a**) the first and (**b**) the second layer (Table 6.13)

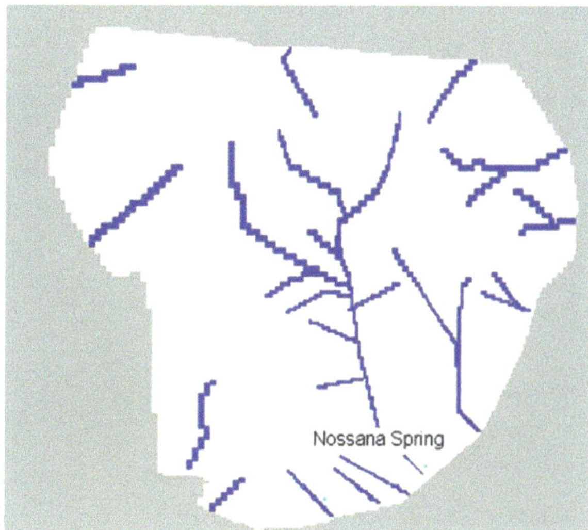

Fig. 6.46 The boundary conditions: no flow cells in *grey*, river in *dark blue* and drain in *blue*

- no flow outside the Nossana hydrogeological catchment area;
- river along the perennial water courses;
- drain for the springs.

The recharge was taken from the historic data of the neighboring rainfall stations. The map of the recharge zones, shown in Fig. 6.47, is based on the pattern of precipitation distribution, geographic features and rock types.

The hydraulic conductivity and porosity values were calibrated with a trial and error technique in transient state, in order to simulate the measured discharge values

Fig. 6.47 The recharge zones (Table 6.14)

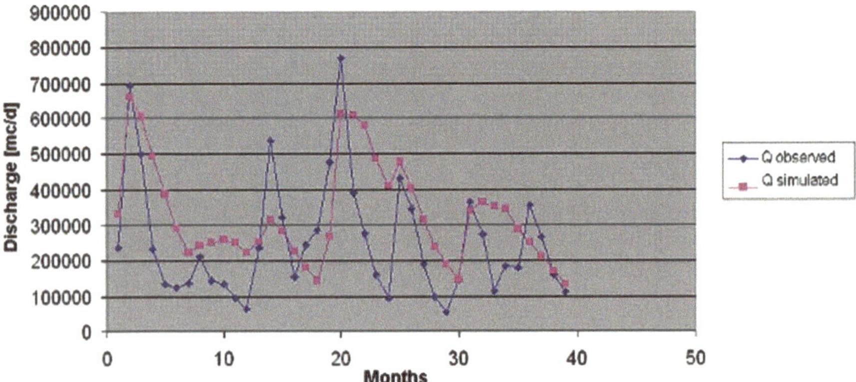

Fig. 6.48 Comparison between the observed discharge of the Nossana Spring and the simulated one

of the Nossana Spring in the last four years (Fig. 6.48). The residual errors of the model are chiefly tied to the great variability of the snow melting. Since there are no data on the hydraulic head of the aquifer, which is known only at the elevation of the spring, no further parameter calibration was possible.

Figure 6.49 shows a piezometrical map achieved in steady state, considering the average value of the rainfall. A small groundwater gradient is evident, typical of karst fields. The flow is directed toward the Nossana Spring where the groundwater rises to the surface, with an average discharge of 3 m^3/s.

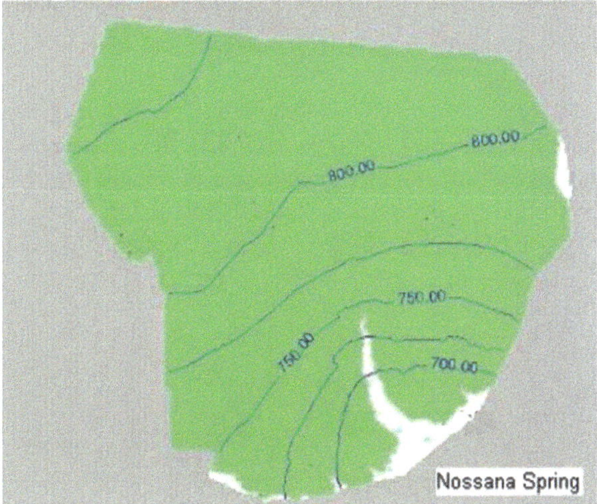

Fig. 6.49 Piezometric map simulated in steady state, with the average rainfall. Groundwater levels are expressed in meters above sea level. The colored cells are the dry superficial cells in the first layer

6.5.3 Factors Involved in the Depletion Curve

The transient state model was used to assess the parameters of the depletion curve of the Nossana Spring. For this aim several transient state simulations were carried out modifying the recharge. For each simulation, the recession trend of the Nossana Spring was analyzed, pointing out its depletion function (Fig. 6.50), which can be expressed through a negative exponential function:

$$Q_t = Q_0 \exp(-\alpha t)$$

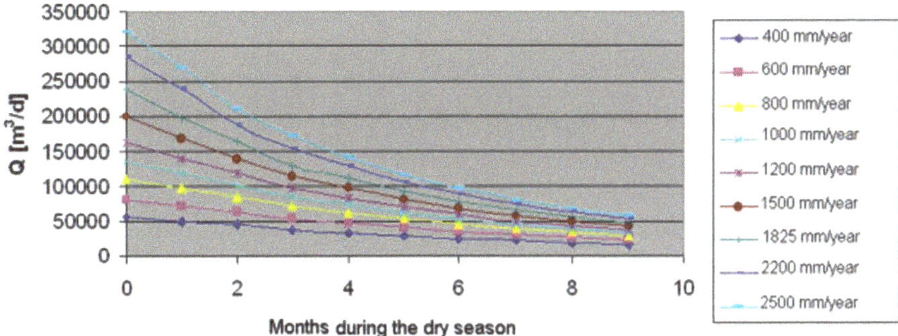

Fig. 6.50 Comparison between the different depletion curves obtained changing the average precipitation of the recharge season

where t is the time since the beginning of the dry season, Q_t and Q_0 are the spring discharges [L^3T^{-1}], respectively, at time t and at initial time, α ($0 \leq \alpha \leq 1$) is the decreasing coefficient [T^{-1}].

The transient state modeling showed that both the initial discharge, Q_0, and the depletion coefficient, α, increase with the precipitation of the recharge season. The results highlight that the function linking the initial discharge, Q_0, and the depletion coefficient, α, to the preceding rainfalls is linear (Fig. 6.51).

Then the depletion curve of the Nossana Spring can be written depending on the rainfalls in the rainy season:

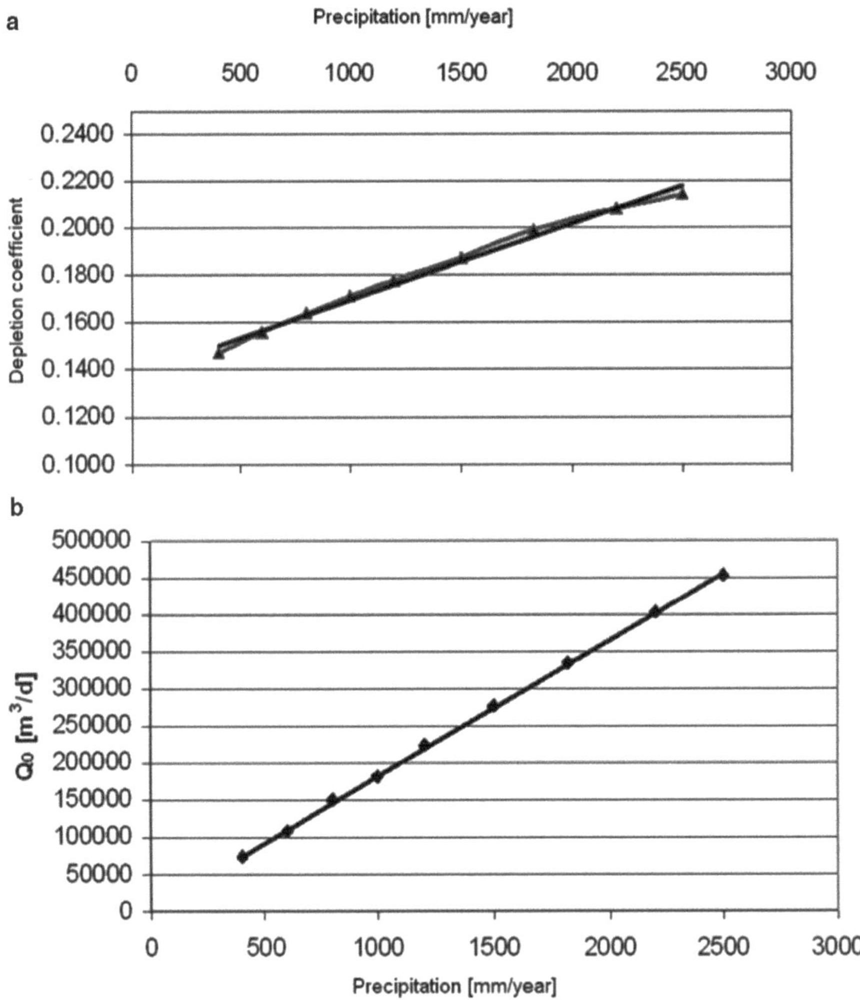

Fig. 6.51 The depletion coefficient, α (**a**) and the initial discharge, Q_0 (**b**) versus the precipitation in the recharge season

Table 6.13 The hydraulic conductivity and porosity values for the different zones (Fig. 6.6), resulting from the calibration

Zone		Hydraulic conductivity [m/s]	Porosity [%]
1		1.0E-03	0.4
2		3.5E-04	0.4
3		2.3E-04	0.3
4		1.2E-04	0.3
5		5.8E-05	0.2
6		2.9E-05	0.2
7		1.4E-05	0.1
8		6.9E-06	0.1

Table 6.14 Average rainfall values and seasonal recharge coefficients for the different recharge zones (Fig. 6.9)

Zone		Average rainfall [mm/year]	Recharge coefficient	
			Winter	Summer
1		1464	0.3	0.3
2		1700	0.3	0.3
3		1800	0.2	0.3
4		1900	0.1	0.3
5		2000	0	0.3
6		2100	0	0.3

$$Q_1 = (181.52r + 2144)\exp[(-9.10^{-4}r - 4.137)t]$$

where Q_t is the spring discharge in m^3/d at time t (expressed in days) and r is the average precipitation (mm/y) in the previous months. Such an evaluation is very useful, since it allows to forecast a crisis event (during the dry season) starting from the monitoring of precipitations in the previous recharge season.

6.5.4 Drying Risk Assessment

Finally, the hazard (meant as the occurrence probability of a crisis event) was assessed. The risk assessment of a spring drying up requires

- the knowledge of the probability distribution of the spring discharge during the depletion period;
- the definition of the critical discharge, $Q_{cr,}$ capable of creating difficulties for the water work.

First of all, as the parameters involved in the depletion function (that is the initial discharge, $Q_{0,}$ and the depletion coefficient, α) depend on random variables (such as the recharge, r, the rock mass permeability, k and its porosity, n), Monte Carlo simulations are required to obtain the probability distribution of the depletion function:

$$Q_t = Q_0(r,k)\exp[-\alpha(k,n)t].$$

For this aim, the synthetic rainfall hyetograms were achieved based on the probability distribution of the rainfalls (Fig. 6.52a); these hyetograms were input in the model to simulate the corresponding discharge hydrograms of the Nossana Spring. So, the probability distribution of the spring discharge at the end of the depletion period was pointed out (Fig. 6.52b). Now the risk can be assessed as

$$R = p[Q_t \leq Q_{cr}],$$

where Q_t is the spring discharge at the end of the dry season (lasting nowadays on average four months). Knowing that the waterworks served by the Nossana Spring have problems for a discharge lesser than 0.5 m^3/s, nowadays the yearly probability of a crisis event is equal to 0.1, but it is likely to become much higher according to the climatic changes.

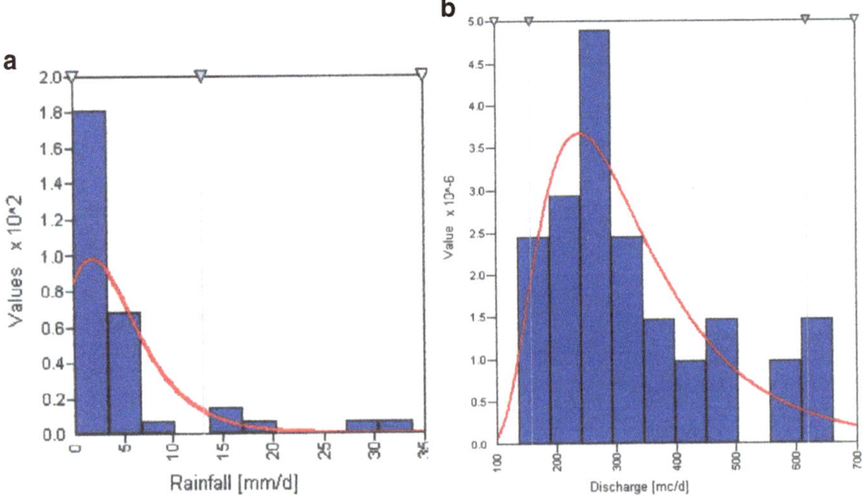

Fig. 6.52 Probability distribution (**a**) of the rainfall (fitted with a Gumbell distribution), (**b**) of the simulated discharge of the Nossana Spring (fitted with a lognormal distribution)

References

Alboin C, Jaffre J, Joly P, Roberts J, Serres C (2002) A comparison of methods for calculating the matrix block source term in a double porosity model for contaminant transport. Comput Geosci 6:523–543

Aler J, Du Mouza J, Arnould M (1996) Measurement of the fragmentation efficiency of rock mass blasting and its mining applications. Int J Rock Mech Min Sci & Geomech Abstr 33:125–140

Allen DM, Macie DC, Wie M (2004) Groundwater and climate change: a sensitivity analysis for the Grand Forks aquifer southern British Columbia. Canada. Hydrogeol J 12(3):270–290

American Society for Testing and Materials (2006) Standard Test Methods for Measurement of Hydraulic Conductivity of Saturated Porous Materials Using a Flexible Wall Permeameter

Anagnostou G (1995) The influence of tunnel tunnelling on the hydraulic head. Int J Numer Anal Meth Geomech 19:725–746

Andersson J, Dverstorp B (1987) Conditional simulations of fluid flow in three-dimensional network of discrete fractures. Water Resour Res 23:1876–1886

Angelini P, Dragoni W (1997) The problem of modelling limestone springs: the case of Bagnara (North Apennines, Italy). Ground Water 35(4):612–618

Baecher GB, Lanney NA, Einstain HH (1977) Statistical description of rock properties and sampling. Proc 18th U.S. Symp on Rock Mech Colorado 5C1.1–5C1.8

Bai M, Elsworth D (1994) Modelling of subsidence and stress-dependent hydraulic conductivity for intact and fractured porous media. Rock Mech Rock Eng 27(4):209–234

Bai M, Elsworth D, Roegiers JC (1997) Triple-porosity analysis of solute transport. J Contam Hydro 28:247–266

Bai M, Meng F, Elsworth D, Roegiers JC (1999) Analysis of stress-dependent permeability in non orthogonal flow and deformation field. Rock Mech Rock Eng 32(3):195–219

Bandis SC, Barton NR, Christianson M (1985) Application of a new numerical model of joint behaviour to rock mechanics problems. Fundamentals of Rock Joints. Proceedings of the International Symposium on Fundamentals of Rock Joints. Bjorkliden, September 1985, Centek, Luleå, Sweden 345–356

Bandis SC, Lumsden AC, Barton NR (1983) Fundamentals of rock joint deformation. Int J Rock Mech Min Sci & Geom Abstr 20:249–268

Barenblatt GE, Zheltov IP, Kochina IN (1960) Basic concepts in the theory of seepage of homogeneous fluids in fissured rocks. J Appl Math and Mech 24:1286–1303

Barla G (2000) Lessons learnt from the excavation of a large diameter TBM tunnel in complex hydrogeological conditions, GeoEng2000. International Conference on Geotechnical and Geological Engineering 938–995

Barton NR (1973) Review of new shear strength of rock and rock joints. Eng Geol 8:287–332

Barton NR, Choubey V (1977) The shear strength of rock joints in theory and practice. Rock Mech 10:1–54

Bear J (1972) Dynamics of Fluids in Porous Media. Elsevier, New York

L. Scesi, P. Gattinoni, *Water Circulation in Rocks*, DOI 10.1007/978-90-481-2417-6_BM2,

Bear J (1993) Modelling flow and contaminant transport in fractured rocks. In: Bear J, Tsang CF, DeMarsily G (eds) Flow and contaminant transport in fractured rock. Academic Press, San Diego
Bear J, Berkowitz B (1987) Groundwater flow and pollution in fractured rock aquifers. In: Nowak P (ed) Development in hydraulic engineering. Elsevier, New York
Beer AJ, Stead D, Coggan JS (2002) Estimation of the Joint Roughness Coefficient (JRC) by visual comparison. Rock Mech Rock Eng 35:65–74
Belem T, Homand-Etienne F, Souley M (2000) Quantitative parameters for rock joint surface roughness. Rock Mech Rock Eng 33(4):217–242
Billaux D (1990) Hydrogéologie des milieux fracturés. Géométrie, connectivité et comportement hydraulique. Document Bureau de Recherches Géologiques et Minières 186
Bini A, Forcella F, Jadoul F, Orombelli G (eds) (2000) Carta Geologica della Provincia di Bergamo alla scala di 1:50000
Birk S, Liedl R, Sauter M (2004) Identification of localised recharge and conduit flow by combined analysis of hydraulic and physico-chemical spring responses (Urenbrunnen, SW-Germany). J Hydrol 286:179–193
Brace WF (1980) Permeability of argillaceous and crystalline rocks. Int J Rock Mech Min Sci & Geomech Abstr 17:241–251
Brouyère S, Carabin G, Dassargues A (2004) Climate change impacts on groundwater resources: modelled deficits in a chalky aquifer, Geer basin. Belgium Hydrogeol J 12:123–134
Brown SR, Kranz RL, Bonner BP (1986) Correlation between the surface of natural rock joints. Geophys Res Lett 13:1430–1433
Burgess A (1977) Groundwater movements around a repository, regional groundwater flow analysis KBS54:03 Kaernbraenslesaekerhet, Stockholm, Sweden
Cacas MC, Ledoux E, DeMarsily G, Tillie B, Barbreau A, Durand E, Feuga B, Peaudecerf P (1990) Modelling fracture flow with a stochastic discrete fracture network: calibration and validation 1. The flow model. Water Resour Res 26:479–789
Cambi C, Dragoni W (2000) Groundwater yield, climatic changes and recharge variability: considerations out of the modelling of a spring in the Umbria-Marche Apennines. Hydrogéologie 4:11–25
Cassan M (1980) Les essais d'eau dans la reconnaissance des sols. Eyrolles, Paris
Catani F (1999) Fractal properties of the Caotic Complex: methods and applications in Tuscany (Italy) Proceedings of the Workshop di Informatica applicata alle scienze della terra, 14–16 settembre, San Sepolcro (AR)
Chardon M (1975) Les Prealpes Lombardes et leurs bordures. Thesis Univo Aix-Marseille
Ciceri E, Gambillara R, Masciocchi N, Monticelli D, Tumiati S (2004) Characterisation of the Eve verda in the St. Marcel valley, Aosta (Italian Western Alps). Int Geol Congr 32(1):359
Citrini D, Noseda G (1987) Idraulica. Casa Editrice Ambrosiana, Milano
Civita M (2005) Idrogeologia applicata e ambientale. Casa Editrice Ambrosiana, Milano
Civita M, De Maio M, Fiorucci A, Pizzo S, Vigna B (2002) Le opere in sotterraneo e il rapporto con l'ambiente: problematiche idrogeologiche. Meccanica e Ingegneria delle Rocce, Torino
Cooper HH, Jacob CE (1946) A generalized graphical method for evaluating formation constants and summarizing well field history. Am Geophys Union Trans 27:526–534
Cotecchia V (1993) Opere in sotterraneo: rapporto con l'ambiente. XVIII Convegno Nazionale di Geotecnica:145–190
Croci A, Francani V, Gattinoni P (2003) Studio idrogeologico del bacino del Torrente Esino. Quaderni di geologia applicata 10(2):148–166
D'Aquino L (2006) Analisi delle prove idrauliche su di un acquifero fratturato attraverso il metodo sperimentale di Kazemi. XXX° Convegno di Idraulica e Costruzioni Idrauliche – IDRA 2006:1–15
De Marsily G (1986) Quantitative hydrogeology: groundwater hydrology for engineers. Academic Press, San Diego
Delleur JW (1999) The handbook of groundwater engineering. CRC Press LLC, Boca Raton

Dematteis A, Kalamaras G, Eusebio A (2001) A systems approach for evaluating springs drawdown due to tunnelling. World Tunnel Congress AITES-ITA 2001 1:257–264
Dershowitz WS (1984) Rock joint systems, Ph.D. Thesis, MIT, Cambridge
Dershowitz WS, Einstein HH (1988) Characterizing rock joint geometry with joint system models. Rock Mech Rock Eng 21(1):21–51
Diodato DM (1994) A compendium of fracture flow codes. Energy System Division, Argonne National Laboratory, Technical Memorandum 96
Domenico PA, Schwartz FW (1990) Physical and chemical hydrogeology. John Wiley & Sons Inc., New York
Drogue G, Pfister L, Leviandier T, Idrissi A, Iffy JF, Matgen P, Humbert J, Hoffmann L (2004) Simulation of the spatio-temporal variability of streamflow response to climate change scenarios in a mesoscale basin. J Hydrol 293:255–269
Dunning CP, Feinstein DT, Hunt RJ, Krohelski JT (2004) Simulation of ground-water flow, surface-eater flow, and a deep sewer tunnel system in the Menomonee Valley, Milwaukee, Wisconsin. USGS Scientific Investigations Report 2004 5031
Dverstorp B, Andersson J (1989) Application of the discrete fracture network concept with field data: possibilities of model calibration and validation. Water Resour Res 25(3):540–550
Edmunds WM, Smedley PL (2000) Residence time indicators in groundwater: the East Midlands Triassic sandstone aquifer. Appl Geochem 15:737–752
El Tani M (2003) Circular tunnel in a semi-infinite aquifer. Tunnelling and Groundwater Space Technology 18:49–55
Elhatip H, Gunay G (1998) Karst hydrogeology of the Kas-Kalkan Springs along the Mediterranean cost of Turkey. Environ Geol 36(1–2):150–158
Federico, F (1984) Il processo di drenaggio da una galleria in avanzamento. R.I.G. 4:191–208
Feng Q, Fardin N, Jing L, Stephansson O (2003) A new method for in-situ non-contact roughness measurement of large rock fracture surfaces. Rock Mech Rock Eng 36(1):3–25
Fiorillo F, Esposito L, Guadagno FM (2007) Analysis and forecast of water resources in an ultra-centenarian spring discharge series from Serino (Southern Italy). J Hydrol 336:125–138
Francani V (1997) Idrogeologia generale e applicata. CittàStudiEdizioni, Milano
Francani V (2002) Gestione dei siti inquinati: dalle indagini alla bonifica. AIGA Bari 21–23 october
Francani V, Fumagalli D, Gattinoni P, Mottini S (2005) Modelli concettuali dinamici per l'analisi del rischio geologico a fini progettuali. Quaderni di Geologia Applicata, 12 (1), Pitagora Editrice (Bologna), pp 35–47
Gangi AF (1978) Variation of whole and fractured porous rock permeability with confining pressures. Int J Rock Mech Min Sci & Geomech Abstr 15
Garbrecht J, Piechota TC (2006) *Climate variations, climate change, and water resources engineering*. American Society of Civil Engineers, Reston
Gates WCB (1997) The Hydro-Potential (HP) value: a rock classification technique for evaluation of the groundwater potential in fractured bedrock. Environ Eng Geosci 3(2):251–267
Gattinoni P, Scesi L, Francani V (2005) Tensore di permeabilità e direzione di flusso preferenziale in un ammasso roccioso fratturato Quaderni di Geologia Applicata 12(1):79–98
Gattinoni P, Papini M, Scesi L (2001) Geological risk in underground excavations. World Tunnel Congress AITES-ITA 2001 vol. 1:309–318
Gattinoni P, Scesi L (2004) Studio degli effetti della rugosità sulla circolazione idrica in un ammasso roccioso. Quaderni di Geologia Applicata 11(2):58–71
Gattinoni P, Scesi L (2007) Roughness control on hydraulic conductivity in fractures. Hydrogeol J, 15:201–211
Gattinoni P, Scesi L (2006) Analisi del rischio idrogeologico nelle gallerie in roccia a media profondità. Gallerie 79 69–79
Gattinoni P, Scesi L, Terrana S (2008) Hydrogeological risk analysis for tunneling in anisotropic rock masses. ITA-AITES World Tunnel Congress "Underground Facilities for Better Environment & Safety", September 19–25 2008 Agra, India

Gentier S, Billaux D (1989) Caracterisation en laboratoire de l'espace fissural d'une fracture. Proc of the International Symp on Rock at Great Depth, Balkema

Gentier S, Billaux D, van Vliet L (1989) Laboratory testing of the voids of a fracture. Rock Mech Rock Eng, 22:149–157

Ghosh A (1990) Role of radial cracks, free face and natural discontinuities of fragmentation from bench blasting. Ph.D. Thesis, University of Arizona

Gianotti F (1999) Geological and geomorphological evidence for the Aosta-Ranzola fault along the Aosta Valley. In: Fission track analysis: theory and application. Mem Sci Geol 51(2): 498–500

Gisotti G, Pazzagli G (2001) L'interazione tra opere in sotterraneo e falde idriche. Un recente caso di studio. In: Teuscher P, Colombo A (eds) AITES-ITA 2001, Progress in tunnelling after 2000, vol. 2. Patron Editore, Bologna

Goodman RE, Moye DG, Van Schalkwyk A, Javandel I (1965) Ground water inflow during tunnel driving. Eng Geol 2:39–56

Grasselli G, Egger P (2003) Constitutive law for the shear strength or rock joints based on three-dimensional surface parameters. Int J Rock Mech Min Sci 40:25–40

Gueguen Y, Dienes J (1989) Transport properties of rocks from statistics and percolations. J Int Assoc Math Geol 21:1–13

Hantush MS (1966) Analysis of data from pumping tests in anisotropic aquifers. J Geophys Res 71(2):421–426

Harbaugh AW, Banta ER, Hill MC and McDonald MG (2000) MODFLOW-2000, the U.S. Geological Survey modular ground-water model – User guide to modularization concepts and the Ground-Water Flow Process. U.S. Geological Survey Open-File Report 00–92

Hsieh PA (2002) Some thoughts on modelling flow in fractured rocks. Int Groundwater Modelling Center Newsletter 20

Huizar AR, Mendez GT, Madrid RR (1998) Patterns of groundwater hydrochemistry in Apan-Tochac sub-basin, Mexico. Hydrol Sci 43(5):669–685

Hung C, Evans DD (1985) A 3-dimensional computer model to simulate fluid flow and contaminant transport through a rock fracture system. NUREG/CR-4042, US Nuclear Regulatory Commission

Huyakorn PS, Lester BH, Faust CR (1983) Finite element techniques for modeling groundwater flow in fractured aquifers. Water Resour Res 19(4):1019–1035

Imray S (1955) Flow of liquids through cracker media. Bull of Water Res. Council Israel 5A(1):84

IPCC (2001) Impact, adaptation and vulnerability. Contribution of the working group II to the third assessment report of the Intergovernmental Panel on Climate Change (IPCC). McCarthy JJ, Canziani OF, Leary NA, Dokken DJ, White KS (eds), Cambridge University Press, Cambridge

Isherwood D (1979) Geoscience database handbook for modelling a nuclear waste repository vol. 1. NUREG/CR-0912VL, UCRL-52719, V1

Itasca (1999) UDEC, User's guide. Itasca Consulting Group Inc., Minneapolis

Ivanova V (1998) Geologic and stochastic modelling of fracture systems in rocks. Ph.D. Thesis, MIT, Cambridge.

Jacob CE, Lohman SW (1952) Nonsteady flow to a well of constant drawdown in an extensive aquifer. Trans Am Geophys Union 33(4):559–569

Jadoul F, Pozzi R, Pestrin S (1985) La sorgente Nossana: inquadramento geologico e idrologico. Riv Mus Sci Nat BG 9:129–140

Jankowski J, Shekarforoush S, Acworth RI (1998) Riverse ion-exchange in a deeply weathered porphyritic dacite fractured aquifer system, Yass, New South Wales, Australia. In: Arehart GB, Hulston JR (eds) Water-rock interaction. Balkema, Rotterdam

Jeong WC, Bruel D, Hicks T (1999) Modelling the influence of fault zone heterogeneity on groundwater flow and radionuclide transport. Proc. of Eurowaste '99, 5th European Commission Conference on Radioactive Waste Management and Disposal and Decommissioning, 15–18 November

Karlsrud K (2001) Control of water leakage when tunnelling under urban areas in the Oslo region. NFF Publication 12

Kawecki MW (2000) Transient flow to a horizontal water well. Ground Water, 38/6:842–850

Kazemi H (1969) Pressure transient analysis of naturally fractured reservoirs with uniform fracture distribution. Soc Pet Eng J 9:451–462

Kendorski FS, Mahtab MA (1976) Fracture patterns and anisotropy of San Manuel Quartz Monzanite. Bull Assoc Eng Geol 13:23–31

Kiraly L (1969) Anisotropie et hétérogénéité de la perméabilité dans les calcaires fissurés. Eclogae Geol Helv 62/2:613–619

Kiraly L, Mathey B, Tripet JP (1971) Fissuration et orientation des cavités souterraines: règion de la Grotte de Milandre (Jura tabulaire) (Cracking and orientation of the subsurface hollows: Grotte de Milandre (French Jura) area). Bull Soc Neuchateloise Sci Nat 94:99–114

Knutsson G, Olofsson B, Cesano D (1996) Prognosis of groundwater inflows and drawdown due to the construction of rock tunnels in heterogeneous media. Res Proj Rep Kungl Tekniska

Kraemer SR, Haitjema HM (1989) Regional modelling of fractured rock aquifers. In: Jousma G et al. (eds) Groundwater contamination: use of models in decision-making. Kluwer Academic Publishers, Dordrecht

Kruseman GP, de Ridder NA (1994) Analysis and evaluation of pumping test data. Second Edition (Completely Revised). ILRI publ. 47

Krutow-Mozgawa A (1988) Métamorphisme dans les sédiments riches en fer ou magnesium de la couverture des ophiolites piémontaises (mine de Servette, Val d'Aoste). Thèse de 3ème cycle, Université P. et M. Curie, Paris VI

La Pointe PL, Wallmann PC, Follin S (1996) Continuum modelling of fractured rock masses: is it usefull? In: Barla G (ed) Eurock 96, Balkema, Rotterdam

Labat D, Mangin A, Ababou R, (2002) Rainfall-runoff relations for karstic springs: multifractal analysis. J Hydrol 256(3–4):176–195

Lambe TW, Whitman RV (1969) Soil mechanics. John Wiley & Sons Inc, New York

Lapcevic PA, Novakowski KS, Cherry JA (1990) The characterization of two discrete horizontal fractures in shale Proc. Technology Transfer Conference. Ontario Ministry of Environment, Toronto

Laroque M, Manton O, Ackerer P, Razack M (1999) Determining karst transmissivities with inverse modeling and an equivalent porous media. Ground Water 37/6: 939–946

Lee CH, Farmer I (1993) Fluid flow in discontinuous rocks. Chapman & Hall, New York

Lee HS, Cho TF (2002) Hydraulic characteristics of rough fractures in linear flow under normal and shear load. Rock Mech Rock Eng 35(4):299–318

Lei S (1999) An analytical solution for steady flow into a tunnel. Ground Water 37(1):23–26

Liu J, Elsworth D, Brady BH, Muhlhaus HB (2000) Strain-dependent fluid flow defined through rock mass classification schemes. Rock Mech Rock Eng 33(2):75–92

Loew S (2002) Groundwater hydraulics and environmental impacts of tunnels in crystalline rocks. Meccanica e Ingegneria delle rocce Torino

Long JCS (1983) A model for steady fluid flow in random three dimensional networks of disk shaped fractures. Water Resour Res 21(8):1105–1115

Long JCS, Witherspoon PA (1985) The relationship of the degree of interconnaction to permeability of fracture networks. J Geophys Res 90(B4):3087–3098

Louis C (1967) Etude des écoulements d'eau dans les roches fissurées et de leur influence sur la stabilité des massifs rocheux. Bull de la Direction des Etudes et Recherches, Serie A Nucléaire, Hydraulique, Termique 3:5–132

Louis C (1974) Introduction à l'hydraulique des roches. Bur Rech Geol Min 4/3:283–356

Ma T, Wang Y, Guo Q (2004) Response of a carbonate aquifer to climate change in northern China: a case study at the Shentou karst springs. J Hydrol 297:274–284

Machado LI, Silva F, Duias R, Laiginhas C (2001) GIS technologies applied to modelling fluid circulation in carbonate massifs – The Estremos anticline (Portugal). Geospatial World, Atlanta

Manev G, Avramova-Tacheva E (1970) On the valuation of strenght and resistance condition of the rocks in natural rock massif. Proc. 2nd Cong. Int. Soc. Rock. Mech., Belgrade, pp 59–64

Martin S, Rebay G, Kienast JR, Mevel C (2008) An eclogitized, oceanic paleohydrothermal field from the St.Marcel Valley (Italian Western Alps). OFIOLITI 33:1–15, ISSN: 0391-2612

Martin S, Tartarotti P (1989) Polyphase HP metamorphism in the ophiolitic glaucophanites of the lower St. Marcel Valley (Aosta, Italy). Ofioliti 14(3):135–156

McCaffrey MA, Adinolfi AM (2003) Evaluating groundwater behavior in jointed rock subject to dewatering. 39th US Rock Mechanics Symposium 21–22 July

McDonald MG, Harbaugh AW (1988) A modular three dimensional finite difference groundwater flow model. U.S. Geological Survey Techniques of Water Resources Investigations, vol. 6

Meenzel L, Burger G (2002) Climate change scenarios and runoff response in the Mulde catchment (southern Elbe, Germany). J Hydrol 267:53–64

Meyer T, Einstein HH (2002) Geologic stochastic modeling and connectivity assessment of fracture systems in the Boston Area. Rock Mech Rock Eng 35(1):23–44

Min KB, Jing L, Stephansson O (2004) Determining the equivalent permeability tensor for fractured rock masses using a stochastic REV approach: method and application to the field data from Sellafield, UK. Hydrogeol J 12(5):497–510

Mojitabai N, Centintas A, Farmer IW, Savely J (1989) In-place and excaveted block size distribution. Proc. 30th US Symposium on Rock Mechanics, West Virginia University

Molinero J, Samper J, Juanes R (2002) Numerical modelling of the transient hydrogeological response produced by tunnel construction in fractured bedrocks. Eng Geol 64:369–386

Moutsopoulos KN, Konstantinidis AA, Meladiotis ID, Tzimopoulos CD, Aifantis EC (2001) Hydraulic and contaminant transport in multiple porosity media. Transport Porous Media 42:265–292

Negi GCS, Joshi V (2004) Rainfall and spring discharge patterns in two small drainage catchments in the Western Himalayan Mountains, India. The Environmentalist 24:19–28

Neuman SP, Walter GR, Bentley HW, Ward JJ, Gonzalez DD (1984) Determination of horizontal anisotropy with three wells. Ground Water 22(1):66–72

Noussan É (1972) Les fontaines colorées. Bull Soc Flore valdôtaine 26:32–35

Oraseanu I, Mather J (2000) Karst hydrogeology and origin of thermal waters in the Codru Moma Mountains, Romania. Hydrogeol J 8(4):379–389

Orlando BM, Burton I (2003) Change: adaptation of water resources management to climate change. IUCN, Gland

Oxtobee JPA, Novakowski KS (2003) Ground water/surface water interaction in a fractured rock aquifer. Ground Water Genuary

Pahl PJ (1981) Estimating the mean length of discontinuity trace. Int J Rock Mech Min Sci & Geom Abstr 18:221–228

Palmstron A (1982) The volumetric joint count: a useful and simple measure of the degree of rock mass jointing Proc 4th Cong Int Assoc of Eng Geol 5:221–228

Papadopulos IS (1965) Nonsteady flow to a well in a finite anisotropic aquifer. International Association of Scientific Hydrogeology Symposium. Am Water Resour Assoc Proc 4:157–168

Papini M, Scesi L, Bianchi B (1994) Studi finalizzati alla previsione delle venute d'acqua in galleria. Costruzioni april

Park K-H, Owatsiriwong A, Lee G-G (2008) Analytical solution for steady-state groundwater inflow into a drained circular tunnel in a semi-infinite aquifer: a revisit. Tunnelling and Underground Space Technology 23:206–209

Picarelli L, Petrazzuoli SM, Warren CD (2002) Interazione tra gallerie e versanti. Meccanica e Ingegneria delle Rocce MIR

Piggott AR (1990) Analytical and experimental studies of rock fracture hydraulics. Ph.D. Thesis at Pennsylvania State University, Pennsylvania

Poulton MM, Mojitabai N, Farmer IW (1990) Scale invariant behavior of massive and fragmented rock. Int J Rock Mech Min Sci & Geomech Abstr 27:219–221

Priest SD, Hudson JA (1976) Discontinuity spacings in rock. Int J Rock Mech Min Sci & Geomech Abstr 13:135–148

Qian J, Zhan H, Wu Y, Li F, Wang J (2006) Fractured-karst spring-flow protections: a case study in Jinan, China Hydrogeol J (2006) 14:1192–1205

Rabinovitch A, Bahat D, Melamed Z (1999) A note on joint spacing. Rock Mech Rock Eng 32:71–75

Ramsay JG, Huber MI (1987) The techniques of modern structural geology. Academic Press, London

Rasmussen TC (1988) Fluid flow and solute transport through three-dimensional networks of variably saturated discrete fractures. Ph.D. Dissertation, University of Arizona

Raven KG, Gale JE (1976) Evaluation of structural and groundwater conditions in underground mines and excavations: subsurface containment of solid radioactive wastes. Geol Survey of Canada Progress Report EMR/JSC-RW

Regione autonoma Valle d'Aosta (2006) Cambiamenti climatici in Valle d'Aosta: opportunità e strategie di risposta. Società metereologica subalpina, Torino

Reuter E, Kopp B, Lemke S (2000) Hallandsas Tunnel. Waterproofing system with a 4 mm thick plastic membrane. Tunnel 6:39–45.

Ribacchi R, Graziani A, Boldini D (2002) Previsione degli afflussi d'acqua in galleria e influenza sull'ambiente. Meccanica e Ingegneria delle rocce, MIR, Torino

Robinson PC (1982) Connectivity of fracture system – A percolation theory approach, Theoretical Physics Division, AERE Arwell. DOE report no. DOE/RW/81.028, march

Romana M (1985) New adjustment ratings for application of Bieniawski classification to slopes. Int. Symp. on the role of rock mechanics ISRM, Zacatecas

Rossi S, Ranfagni L, Biancalani P, Calzolai L (2001) Geological and hydrogeological analysis in large scale tunnelling and impact forecasting on groundwater resources: Bologna-Florence High Speed Railway (Italy). World Tunnel Congress AITES-ITA 2001 1:649–656

Rouleau A, Gale JE (1985) Statistical characterisation of the fracture system in the Stripa Granite, Sweden. Int J Rock Mech Min Sci & Geomech Abstr 22:353–367

Rowe RK, Booker JR (1990) Contaminant migration in a regular two or three dimensional fractured network: reactive contaminants. Int J Numer Anal Methods Geomech 14:401–425

Samardzioska T, Popov V (2005) Numerical comparison of the equivalent continuum, non homogeneous and dual porosity models for flow and transport in fractured porous media. Adv Water Resour 28(3):235–255

Sauro U (1993) Human impact on the karst of the Venetian forealps, Italy. Environ Geol 21(3): 115–121

Scanlon BR, Mace RE, Barret ME, Smith B (2003) Can we simulate regional groundwater flow in a karst system using equivalent porous media models? Case study, Barton Springs Edwards aquifer, USA. J Hydrol 276:137–158

Scesi L (1993) Circolazione idrica in ammassi rocciosi cristallini e opere in sotterraneo Le Strade marzo-aprile

Scesi L, Papini M, Gattinoni P (2006) Geologia Applicata. Il rilevamento geologico-tecnico. Casa Editrice Ambrosiana, Milano

Scesi L, Saibene L (1989) Verifica sperimentale della validità del rilevamento geologico-strutturale per fini applicativi. Le Strade XCI:1262

Shante VKS, Kirkpatrick S (1971) Introduction to percolation theory. Adv Phys 20:325–357

Shapiro A, Andersson J (1983) Steady state fluid response in fractured rock: a boundary element solution for a coupled, discrete fracture continuum model. Water Resour Res 19(4):959–969

Snow DT (1969) Anisotropic permeability of fractured media. Water Resour Res 5:1273–1289

Snow DT (1970) The frequency and apertures of fractured rock. Int J Rock Mech Min Sci 7:23–40

St. Martin de La Motte (Comte de) (1784–1785) Sur la fontaine verte de Saint Marcel dans la Vallée d'Aoste. *Mémoires de Mathématiques et de Physique tirés des régistres de l'Académie Royale des Sciences*, Turin.

Sudicky EA (1990) The Laplace transform Galerkin technique for efficient time-continuous solution of solute transport in double-porosity media. Geoderma 46:209–232

Tartarotti P (1988) Le ofioliti piemontesi nella media e bassa valle di St. Marcel (Aosta) Ph.D. Thesis, Università di Padova

Tartarotti P, Martin S, Polino R (1986) Geological data about the ophiolitic sequences in the St. Marcel valley (Aosta Valley). Ofioliti 11:343–346

Therrien R (1992) Three dimensional analysis of variably-satured flow and solute transport in discretely-fractured porous media, Ph.D. Thesis, University of Waterloo

Thiel K (1989) Rock mechanics in hydroengineering. Elsevier, New York

Tumiati S, Godard G, Masciocchi N, Martin S, Monticelli D (2007) Environmental factors controlling the precipitation of Cu-bearing hydrotalcite-like compounds from mine waters. The case of the "Eve verda" spring (Aosta Valley, Italy). Eur J Mineral 20(1):73–94

Turcotte DL (1986) Fractal and fragmentation. J Geophys Res 92:1921–1926

Tweed SO, Weaver TR, Cartwright I (2005) Distinguishing groundwater flow paths in different fractured-rock aquifers using groundwater chemistry: dandenong ranger, southeast Australia. Hydrogeol J 13:771–786

Veneziano D (1978) Probabilistic model of joints in rock. unpublished manuscript, MIT, Cambridge

Vigna B (2001) Gli acquiferi carsici. Quaderni didattici SSI 12:3–48

Villaescusa E, (1993) Statistical modelling of rock jointing. In: Li & Lo (eds) Probabilistic methods in geotechnical engineering. Balkema, Rotterdam

Walsh JB (1981) Effects of pore pressure and confining pressure on fracture permeability. Int J Rock Mech Min Sci & Geomech Abstr 18:429–435

Walsh JB, Brace WF (1984) The effects of pressure on porosity and the transport properties of rock. J Geophyf Res 80:9425–9431

Warren JE, Root PJ (1963) The behavior of naturally fractured reservoirs. Soc Pet Eng J Trans ASME 228:245–255

Weeks EP (1969) Determining the ratio of horizontal to vertical permeability by aquifer-test analysis. Water Resour Res 5:196–214

Witherspoon PA, Wang JSY, Iwai K, Gale JE (1980) Validity of cubic law for fluid flow in a deformable rock fracture. Water Resour Res 16(6):1016–1024

Woldeamlak ST, Batelaan O, De Smedt F (2007) Effects of climate change on the groundwater system in the Grote-Nete catchment, Belgium. Hydrogeol J 15:891–901

Wolkersdorfer C, Bowell R (2005) Contemporary reviews of mine water studies in Europe, Part 3. Mine Water Environ 24:58–76

Yang ZY, Lo SC, Di CC (2001) Reassessing the joint roughness coefficient (JRC) estimation using K2. Rock Mech Rock Eng 34(3):243–251

Index

SPRINGER NATURE

GPSR Compliance

The European Union's (EU) General Product Safety Regulation (GPSR) is a set of rules that requires consumer products to be safe and our obligations to ensure this.

If you have any concerns about our products, you can contact us on ProductSafety@springernature.com

In case Publisher is established outside the EU, the EU authorized representative is:

Springer Nature Customer Service Center GmbH
Europaplatz 3
69115 Heidelberg, Germany

Zeitfracht Medien GmbH
Ferdinand-Jühlke-Straße 7
99095 Erfurt, Deutschland
produktsicherheit@kolibri360.de